INSECT

INSECTRONICS

BUILD YOUR OWN WALKING ROBOT

KARL WILLIAMS

McGraw-Hill

New York Chicago San Francisco Lisbon London Madrid
Mexico City Milan New Delhi San Juan Seoul
Singapore Sydney Toronto

The McGraw-Hill Companies

Cataloging-in-Publication Data is on file with the Library of Congress

1 2 3 4 5 6 7 8 9 0 DOC/DOC 0 8 7 6 5 4 3 2

ISBN 0-07-141241-7

The sponsoring editor for this book was Judy Bass and the production supervisor was Pamela A. Pelton. It was set in Times New Roman by Patricia Wallenburg.

Printed and bound by RR Donnelley.

This book is printed on recycled, acid-free paper containing a minimum of 50% recycled, de-inked fiber.

Dedicated to the memory of my sister
Margaret Frances Williams 1965–2000

CONTENTS

INTRODUCTION

Walking robots are one of the most interesting results of today's advanced technology. These walking machines have introduced an exciting new element to robotics and they have piqued the interest of many. This technology is being driven forward by independent researchers, roboticists, and hobbyists. It is a unique area of study because it encompasses many different disciplines such as electronics, computer science, mechanical design, control systems, programming, and biology; each of which requires a certain expertise to master. The dream of creating artificial life is an old one that dates back to the ancient Greeks, Arabs, Leonardo da Vinci, Pierre Jaquet-Droz, and Nikola Tesla. The robot fascinates man because it is a machine that closely resembles life and man himself. In our lifetime we are witnessing the emergence of intelligent machines that can interact with the world almost as well as we do. The goal of this book is not only to explain in detail how to build a walking robotic life-form that can sense and respond to its environment but also as a starting point for the more advanced development of artificially intelligent robotic entities. I hope that you have as much fun building and programming the robot in this book as I did. Build the robot as it is or take what you want from the designs and incorporate them into your own robotic creations. After building your own artificially intelligent insect-like robot you may begin to question the very nature of intelligence and life itself.

ACKNOWLEDGMENTS

I would like to acknowledge the following people for their support throughout the writing of this book. Thanks to my parents Gord and Ruth Williams for supporting my interest in robotics, electronics, and computers throughout the years. My brothers, Doug Williams and Geoff Williams, who have helped inspire and build many crazy electronic, computer, and robotic devices. Thanks to Laurie Borowski for her love, patience, and suggestions during the writing of this book. Thanks to Judy Bass at McGraw-Hill for doing such a great job and making this book happen. Thanks to Patricia Wallenburg for doing a fantastic job of putting the book together and adding the flip book animations. Thanks to the core group of people who regularly visit the hardware lab at Mitra Imaging where the talk often focuses on robotics and the fantastic future that we will make possible through technology: James Vanderleeuw, Paul Steinbach, Doug Williams, Darryl Archer, Stacey Dineen, Michael Hiemstra, Tom Cloutier, Charles Cummins, Dan Dubois, John Lammers, Raymond Pau, Clark MacDonald, Sachin Rao, Sharilyn Allen, Denise Williams, Steve Hendrikse, Jon Haywood, and Dave Huson. To my friends who have helped in one way or another: Jason Jackson, Kenn Booty, David Greene, Roland Hofer, Kelly Tynkkynen, Ramone Seward, JoAnna Kleuskens, Richard Peers, Richard Mecrate, Brenni Walls, Jean Cockburn, Patti Ramseyer, Bob Hanley, Brent Germin, Jef Horst, Mary Dever, Dave Gard, and Peter Conrad. Thanks to Eric Peterson and Larry Williamson, co-founders of Mitra Imaging. Thanks to Myke Predko, Sharman Crockett, Steve Frederick, Terri Zuccherato, Sameer Siddiqi, Roger Skubowius, Rodi Snow, Cindy Long, Rob Black, Chris French, Mark Penner, Ros Mountford, Angela Wolfe, David Borowski, Rado Ristovic, Bert Wikkerink, Paul Carvalho, Beth Whittle, Brenda Rankin, Darren Tarachan, Dave Gackstetter, Steve Rankin, Glenn Poole, Chris Meidell, Jack Kesselman, Jack Van Ham, Dan Keithlin, Jim Edwards, Kristi Fox, Michelle MacLean, and Sudip Biswas who have all provided suggestions and encouragement. Thanks to James Burbidge, Eric Squire, Dave Hildebrandt, and Rob Lee for the great times when we built robots together during the early 1980s.

1

A BRIEF HISTORY OF WALKING MACHINES

The first walking machines in history were mechanical toys like the ones pictured in Figure 1.1 and Figure 1.2. The toys usually operated from a rotary power source such as a windup spring and clockwork, driving cranks, or cams to which the legs were attached. The legs most often executed a fixed cycle to achieve forward motion. These toys could be considered the proof of concept devices that instilled the idea that larger walking machines could be possible.

The invention of the internal combustion engine was the main event that made practical walking machines possible. The earliest attempt to construct a walking machine was made in Britain in 1940 by A.C. Hutchison and F.S. Smith. Hutchison and Smith built a machine with four independently controlled legs that walked using the quadruped crawl gait. The robot was intended for use as a rough terrain armored vehicle in the 1000-ton class. The device used a pair of hydraulic actuators with a rolling thigh joint, which acted as a kind of inverted wheel, and a telescoping leg. The control mechanism was a series of levers connected to the hands and legs of the operator in a feedback loop with the mechanical legs. This same concept was used by General Electric for its walking truck in the 1960s. A small 60-cm high model was built. The eight joints were controlled by flexible cables, which led to a console where the operator sat. The machine was capable of climbing over a small pile of books. The project was canceled in 1940 when the U.K. war department pulled funding in favor of the well-established tracked tank.

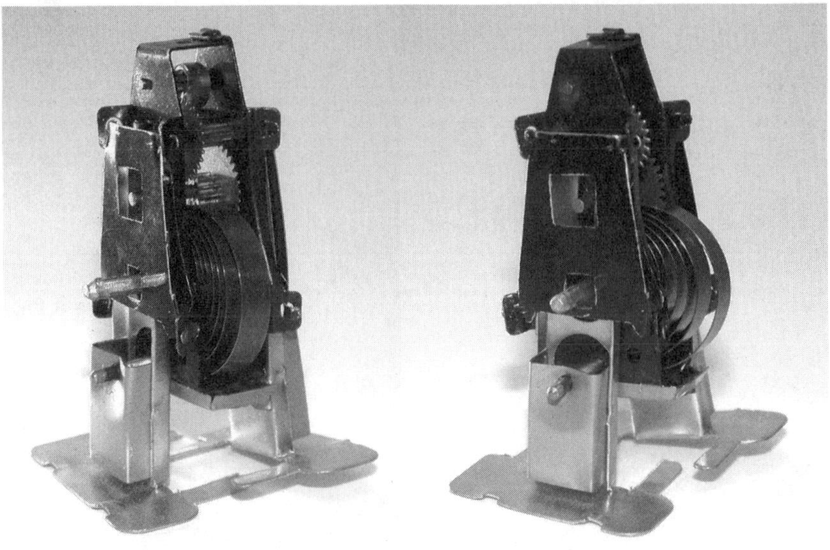

FIGURE 1.1 **1950s walking robot with a windup spring mechanism.**

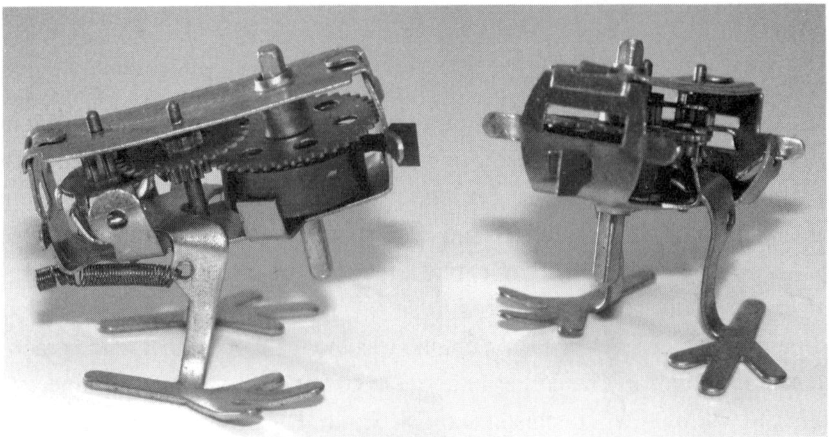

FIGURE 1.2 **1950s hopping robot with a windup spring mechanism.**

The main course of modern legged vehicle research began with the University of Michigan and the U.S. Army Tank-Automotive Center (ATAC) at Warren, Michigan. The U.S. military has maintained an interest in machines for rough ground transport since the Second World War. The Defense Advanced Research Projects Agency (DARPA) has funded much of the research into walking machines. The group started with M. G. Bekker, who founded the ATAC's land-locomotion laboratory in the early

1950s. R.A. Liston continued the work of the laboratory and wrote a much-referred to paper on walking machines in *Journal of Terramechanics* (1964). Another associate was R. K. Bernhard of Rutgers University, who in 1957 began an investigation of gaits for legged vehicles. In a 1957 study, he examined several types of levered vehicles, including leaping, galloping, bouncing and running machines. His focus was on vehicle speed, but these schemes were difficult because of control and vibration problems. The most viable seemed to be running.

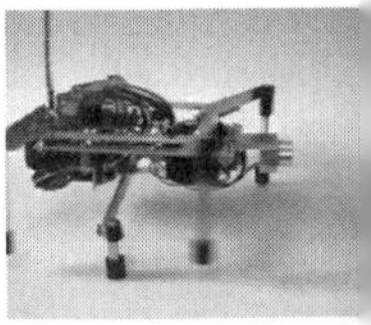

Soon after, J.E. Shigley of the University of Michigan came up with a study that established a set of criteria for the ideal walking machine. Shigley designed and built a machine that was intended to fulfill most of these requirements. The machine had problems stemming from the use of non-circular mechanical linkages and the project was abandoned in favor of an hydraulic system. At the time there was no adequate way of controlling such a system over rough ground with obstacles, so this project was also abandoned.

In 1962, the land locomotion laboratory started work on a new project with General Electric because of its success with building remote manipulators using force feedback from the operator's arms. The most well known of these "cybernetic anthropomorphous machines" as General Electric called them was the Handiman. A photograph published in the media of the Handiman being used to whirl a hula-hoop made it a popular machine. H. Aurand, of General Electric, thought that a bipedal walking machine could be built using the same force feedback servo principle. It was not considered to be an exoskeleton because it was much larger than a human. General Electric decided that it would be better to build a quadruped. The machine that was built, called the "walking truck," was designed by R.S. Mosher of General Electric. Each leg of the walking truck was servo-controlled by the human driver. The front legs of the machine were controlled by the human's arms and the back legs followed the movements of the operator's own legs. The machine weighed 1400 kg, was over 3 meters long, and the hydraulic power was generated by pumps connected to a 90-horsepower engine. The quadruped functioned well, but was very demanding on the driver, who could only manage to control it for a few minutes at a time. This walking machine was important because it proved that the mechanical engineering of legged machines was possible. The fact that it was highly publicized probably stimulated other walking robotic research.

Aerojet General Corporation, on behalf of the U.S. Army and NASA, undertook the next series of walking machine developments. NASA and the U.S. army were interested in the problems of mobility on the moon and other planets. The possible uses of such machines were planetary exploration, a walking chair for the disabled and military transport. Different versions of the basic vehicle had six and eight legs and executed an alternating tripod and an alternating tetrapod gait. The idea was that a train of these vehicles would link together and walk behind a soldier who would control the steering.

This configuration was known as the "Iron Mule Train." This line of research was also abandoned because the performance was not good enough for the infantry supply application. This was probably the last attempt to build purely mechanically coordinated walking machines without computerized or electronic control systems.

Computer Control

Since the 1960s, almost all legged machines have used some kind of computer control. Stored-program control is needed because it provides the flexibility to modify the coordination of the various legs to achieve a stable walking gait.

A brief timeline of modern walking machine research follows:

- **1965**—The Phony Pony. R. McGhee decided that machines like the General Electric walking truck could be controlled using a stepping cycle generated by a computer or electronic logic. The Phony Pony was built by McGhee and A.A. Frank in 1966 and was sometimes referred to as the "Californian Horse." It was a quadruped with two degrees of freedom per leg.
- **1973**—The Waseda University Bipeds. One of the most popular robots at the time was the Wabot because of its resemblance to a human. It first walked in 1973 and was statically stable at all times by keeping its center of mass above one of its large feet at all times.
- **1974**—An electric hexapod is built by Petternella of the University of Rome. This machine was unusual because of the use of telescopic leg joints for the vertical motion and pivoting at the hips. Its shape resembled a table.
- **1977**—The Ohio State University (OSU) Hexapod. This six-legged machine weighed about 100 kg and took its first steps in 1977. Each of the six legs had three joints and was originally controlled by a PDP-11.
- **1977**—Kyushu Institute of Technology quadruped. This pneumatic quadruped overcame the difficulty of proportional control of pneumatics by using a chain of three actuators whose strokes could be combined mechanically.
- **1979**—The eight-legged machine ReCUS (Remotely Controlled Underwater Surveyor) was built by the Technical Research Center Komatsu Ltd. Japan. The machine was 8 meters long, 5 meters wide, 6 meters high and weighed 29 tons in air.
- **1983**—The Carnegie Mellon University (CMU) Hexapod was built between 1980 and 1983 by Southerland and Sproull. The machine was 2.4 m long and reached a maximum speed of 0.11 m/s. This machine was significant because it was the first man-carrying computer-controlled walking machine.

- **1983**—The Odetics six-legged robot Odex I. This machine had a startling appearance with its six legs disposed symmetrically about the vertical axis.
- **1984**—The Ohio State University Adaptive Suspension Vehicle. This six-legged walking machine was powered by a .09-liter motorcycle engine and intended for transportation across rough terrain.

- **1985**—TITAN IV. This four-legged machine was displayed at the Government Pavilion of the Science Exhibition at Tsukuba, Japan in 1985. The name is an acronym from "Tokyo Institute of Technology, Aruku Norimono (walking vehicle)." This machine realized static and dynamic fusion gaits in which gaits were automatically switched among a static state, a crawl gait, a dynamic state, and a trot gait, in which it walked using two diagonal legs alternately. TITAN IV walked at a velocity of 40 cm/sec.
- **1989**—The machine Aquarobot was developed and constructed from 1985 to 1989, at the Robotics Laboratory, Port Harbor Research Institute in Japan.
- **1989**—Genghis. This six-legged machine was developed by Grinnell More, Rodney Brooks, and Colin Angle at Massachusetts Institute of Technology.

From this point in time onward, walking robots have started to appear in all the robotics labs throughout the world. Independent researchers and hobbyists have done much interesting work as well. Figure 1.3 shows a 2-legged autonomous walking robot built by Williams in 1999. It would be impossible to list all the projects that now exist. For up-to-date developments in walking machine research, it is best to search the Internet to find information. One such Internet site that lists new walking machine developments on a monthly basis is called *The Walking Machine Catalogue* and is moderated by Karsten Berns. The URL is http://www.fzi.de/divisions/ipt/WMC/preface/preface.html.

Summary

Studying the history of walking machines and robotics gives the experimenter a sense of what has been accomplished and the creative potential that still exists for future robots. We can learn from past walking machine projects by observing the techniques that worked and the ones that didn't, in order to increase the likelihood of constructing successful walking robots.

In the next chapter, we will discuss the various walking gaits used by legged insects and animals to achieve locomotion, and how this information can be applied to walking machines.

FIGURE 1.3 A two-legged walking robot named Lima1.

2

WALKING GAITS

A robot's gait is the sequence of leg movements it uses to move from one place to another. Each leg movement is broken down into step cycles, where a cycle is complete when the leg configuration is in the same position as it was when the cycle was initiated. A walking gait is a repetitive pattern of foot placement causing a regular progression forwards or backwards. Animals and insects choose various gaits depending on terrain and desired speed. With six legs there are many gaits, but the two most popular are the *alternating tripod gait* and the *wave gait*. Six legs allow the robot to have three legs on the ground at all times, making it a "stable tripod." Knowledge of gaits provides the programmer with a base from which control algorithms or sequences for walking machines can be written. It can also be helpful when designing a walking machine. A legged robot can be defined as a servomechanism with many degrees of freedom. The robot's legs are connected by movable joints and actuators of some sort to power the joints. Control of the actuators' movements, varying over time, will result in sustained stable motion of the machine in a specified direction. Sustained stable motion consists of several objectives, such as stability, maintaining body orientation, control of forward velocity, the ability to turn and reverse, the ability to walk on rough ground and the ability to avoid obstacles.

Stability

One of the main objectives with legged locomotion is to achieve stability. All legged animals and machines face the potentially dangerous problem of falling over, because any variations in the slope of the ground could result in unstable states. Studies of walking have indicated that controlled falling may be a key part of walking. With locomotion, stability is more a matter of achieving a stable cycle where parts of the cycle itself may be quite unstable. A six-legged robot is stable when at least three of its legs are touching the ground, forming a tripod. This is one of the reasons that hexapods are popular.

During locomotion, the simplest form of stability is to pass without a break from one stable state to another. Most walking machines pass through stages in the locomotion cycle which are not stable, and the machines fall temporarily. If the gait does not contain some phases which are stable, the machine will not be able to stop moving without falling over unless it changes its gait. Most animals change from stable gaits at low speed to more unstable gaits at higher speeds where balance comes into play. Since it is difficult to achieve dynamic balance with a walking machine, most walking machines have been designed to balance statically. The most common way to do this is to use six legs and move them in triplets so that three legs always support the robot.

Adaptability

When a walking robot must traverse ground that is not smooth and level, several control problems may arise. These problems include navigation where the robot must choose a route that will get it to a specified location. The second problem is with path selection. Given a specified location, the robot must choose the details of the route to minimize the problems of slope, roughness, and obstacle avoidance. The third problem is with terrain adaptation and obstacle crossing.

TRIPOD GAIT

With the alternating tripod gait, the legs are divided into two sets of three legs. Each set is comprised of the front and rear legs on one side, along with the middle leg on the opposite side. A forward step half cycle starts with one set of legs lifting and moving to a forward position. At the same time the motors of the legs in contact with the ground swing backwards the same amount, moving the body of the robot forward. The lifted legs are then lowered, completing the half cycle. The lifting and support sets are interchanged and the second half of the cycle is completed, as shown in Figure 2.1.

This is the fastest stable gait. During the step, the weight of the robot is only supported on three legs. If one fails to find firm footing, the robot will fall. The robot discussed in this book walks using a simplified version of the alternating tripod gait, as shown in Figure 2.1. The middle legs lift the body and are used as a swivel, but they never actually move in a forward or reverse direction. The exact forward walking sequence to control the robot in this book is detailed in Figure 2.2. The walking sequence that will be used to rotate the robot to the left or right is shown in Figure 2.3.

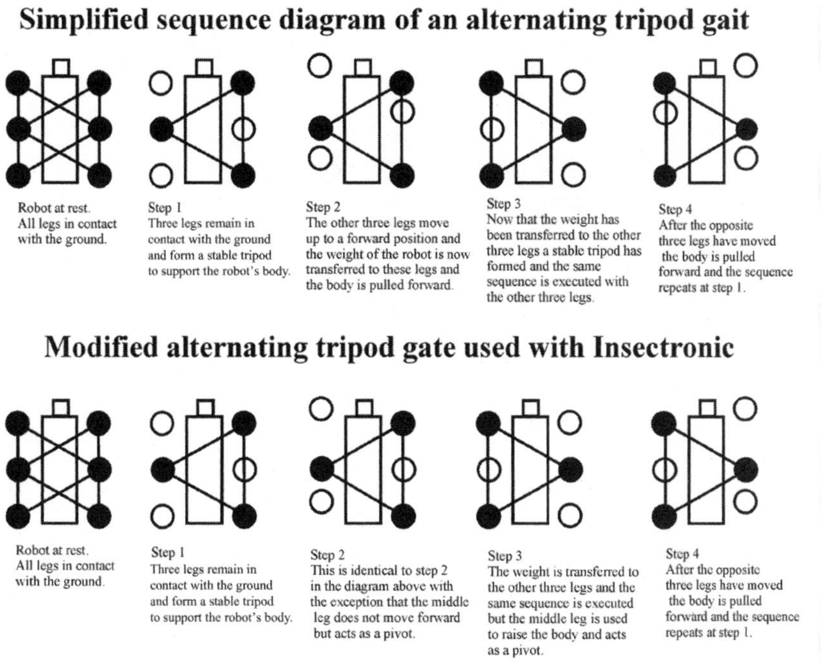

Simplified sequence diagram of an alternating tripod gait
Robot at rest. All legs in contact with the ground.
Step 1 Three legs remain in contact with the ground and form a stable tripod to support the robot's body.
Step 2 The other three legs move up to a forward position and the weight of the robot is now transferred to these legs and the body is pulled forward.
Step 3 Now that the weight has been transferred to the other three legs a stable tripod has formed and the same sequence is executed with the other three legs.
Step 4 After the opposite three legs have moved the body is pulled forward and the sequence repeats at step 1.
Modified alternating tripod gate used with Insectronic
Robot at rest. All legs in contact with the ground.
Step 1 Three legs remain in contact with the ground and form a stable tripod to support the robot's body.
Step 2 This is identical to step 2 in the diagram above with the exception that the middle leg does not move forward but acts as a pivot.
Step 3 The weight is transferred to the other three legs and the same sequence is executed but the middle leg is used to raise the body and acts as a pivot.
Step 4 After the opposite three legs have moved the body is pulled forward and the sequence repeats at step 1.

FIGURE 2.1 Tripod gaits.

In Chapter 10 when the actual walking gaits are programmed, the robot will also use a gait sequence that will allow it to move in a forward or reverse direction while turning to the left or right at the same time.

WAVE GAIT

A more stable, but slower choice is the wave gait, where only one leg is lifted at a time. Starting with a back leg, it is lifted and moved forward. All the other grounded legs are moved back by one-sixth of the amount of the forward motion. The lifted leg is then lowered, and the process repeated for the next leg on the same side. Once the

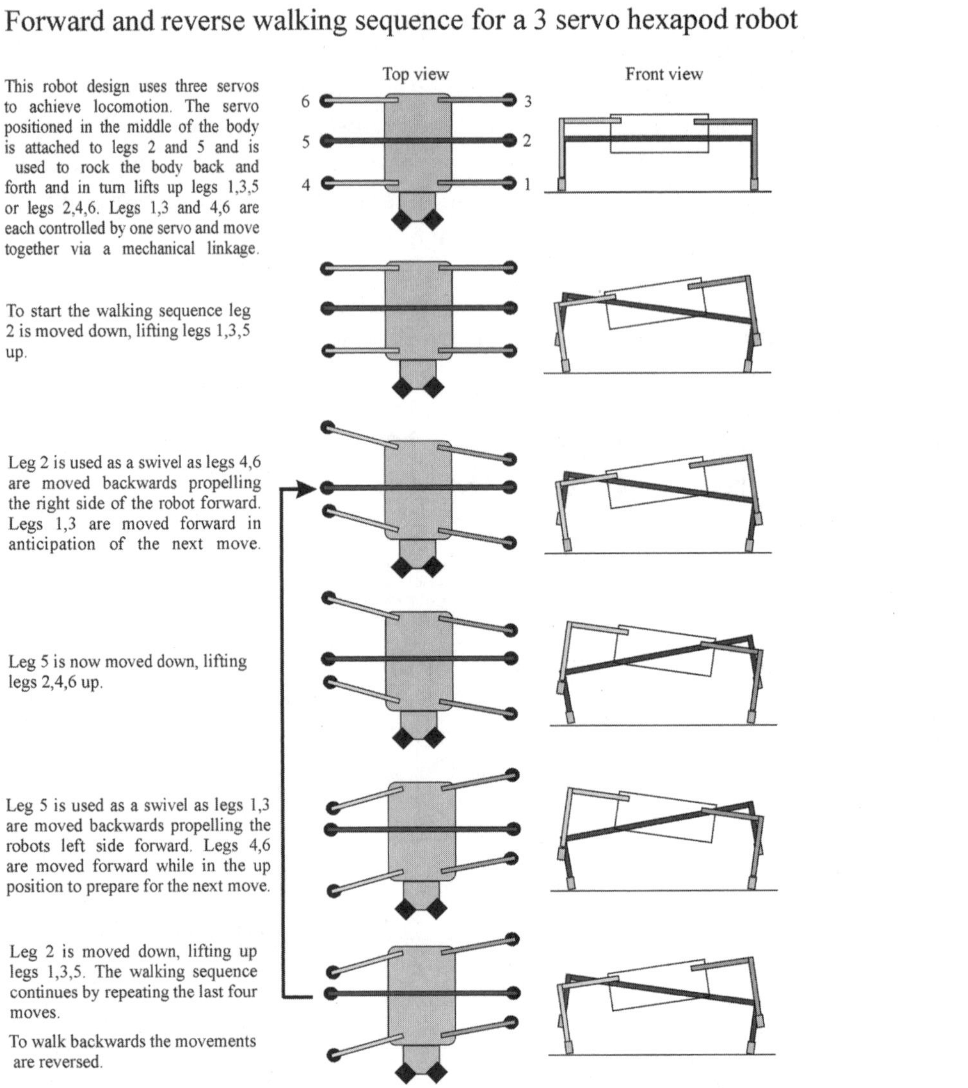
Forward and reverse walking sequence for a 3 servo hexapod robot
Top view
Front view
6
5
4
3
2
1
This robot design uses three servos to achieve locomotion. The servo positioned in the middle of the body is attached to legs 2 and 5 and is used to rock the body back and forth and in turn lifts up legs 1,3,5 or legs 2,4,6. Legs 1,3 and 4,6 are each controlled by one servo and move together via a mechanical linkage.
To start the walking sequence leg 2 is moved down, lifting legs 1,3,5 up.
Leg 2 is used as a swivel as legs 4,6 are moved backwards propelling the right side of the robot forward. Legs 1,3 are moved forward in anticipation of the next move.
Leg 5 is now moved down, lifting legs 2,4,6 up.
Leg 5 is used as a swivel as legs 1,3 are moved backwards propelling the robots left side forward. Legs 4,6 are moved forward while in the up position to prepare for the next move.
Leg 2 is moved down, lifting up legs 1,3,5. The walking sequence continues by repeating the last four moves.
To walk backwards the movements are reversed.

FIGURE 2.2 Forward and reverse walking sequence for Insectronic.

front leg has been moved, the procedure is repeated for the other side. The attraction of this gait is that there are always at least five legs supporting the weight of the robot. But the forward speed is only one-sixth of the alternating tripod gait.

DEGREES OF FREEDOM PER LEG

On each side of the robot that we will be building, the front and rear legs are linked so that one servo can be used to move the two legs forwards and backwards. The third

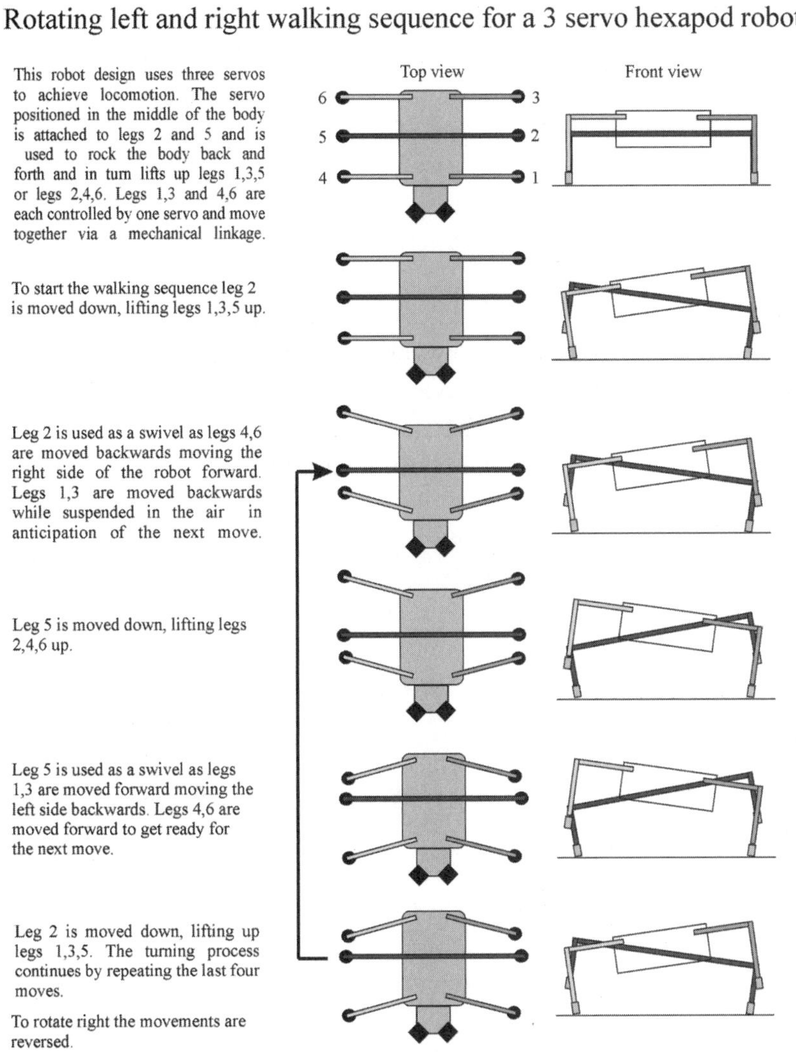

Rotating left and right walking sequence for a 3 servo hexapod robot
Top view
Front view
6
5
4
3
2
1
This robot design uses three servos to achieve locomotion. The servo positioned in the middle of the body is attached to legs 2 and 5 and is used to rock the body back and forth and in turn lifts up legs 1,3,5 or legs 2,4,6. Legs 1,3 and 4,6 are each controlled by one servo and move together via a mechanical linkage.
To start the walking sequence leg 2 is moved down, lifting legs 1,3,5 up.
Leg 2 is used as a swivel as legs 4,6 are moved backwards moving the right side of the robot forward. Legs 1,3 are moved backwards while suspended in the air in anticipation of the next move.
Leg 5 is moved down, lifting legs 2,4,6 up.
Leg 5 is used as a swivel as legs 1,3 are moved forward moving the left side backwards. Legs 4,6 are moved forward to get ready for the next move.
Leg 2 is moved down, lifting up legs 1,3,5. The turning process continues by repeating the last four moves.
To rotate right the movements are reversed.

FIGURE 2.3 Left and right turning sequence for Insectronic.

servo is connected to the middle pair of legs so that rotation of the servo pushes one leg down while lifting the other. The fixed alternating tripod gait is a swivel motion on the down middle leg, driven by the front and rear legs on the opposite side. Although Insectronic can manage to climb over small objects, the lack of independent leg control makes intelligent obstacle climbing difficult.

A leg only needs to have two degrees of freedom (forwards and backwards, up and down) to move. To control the lateral placement of the foot, an extra knee must be added. The consequences of moving from a two-degrees-of-freedom leg to a three-degrees-of-freedom leg are not small. There is the added cost and weight of the extra knee motors, drive electronics, and batteries needed for power. Input is needed from extra sensors to choose the best lateral contact position of the foot. More computing power is needed to service the extra motors, process the new sensor input, and run the more complicated algorithms for three-degrees-of-freedom legs.

Summary

Now that we have covered the basics of walking gaits for six-legged insects, this information can be used in later chapters when developing control programs to sequence the robot's legs to achieve stable walking gaits. In the next chapter, the various tools, test equipment, and materials that are required to build a walking robotic life-form will be discussed.

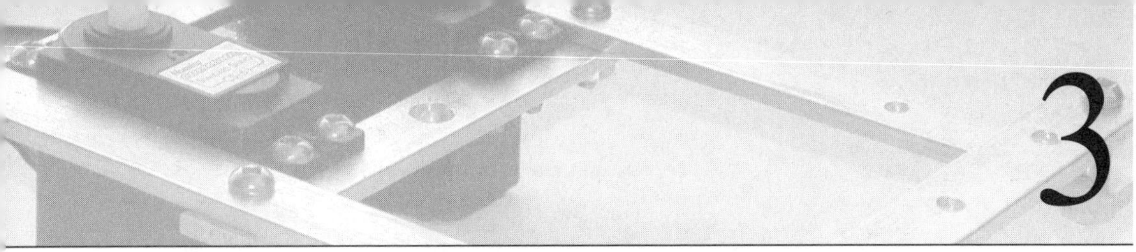

3

TOOLS, TEST EQUIPMENT, AND MATERIALS

During the construction phase of building the robot, there are a number of tools that will be required. You will need a workbench or sturdy table in an area with good lighting. Try to keep your work area clean and free of clutter (which is hard for me to do). The first tool that will be used is the hacksaw. The hacksaw is designed to cut metal and hard plastics. When using the hacksaw to make straight cuts, it is a good idea to use a miter box. Figure 3.1 shows the hacksaw (labeled L) and the miter box (K).

If you have a little extra money and think that you will be building a lot of robots, then what you really need is a band saw fitted with a metal cutting blade. The band saw shown in Figure 3.2 is a 9-inch saw, meaning that the saw can cut pieces up to a maximum length of 9 inches. This is perfect for building smaller robots like the one detailed in this book. With the metal cutting band saw, pieces of aluminum can be cut fast and with greater accuracy than with a hacksaw.

An important piece of equipment that will be needed in your workshop is a vise, like the one shown in Figure 3.3. The vise will be needed quite often when cutting, drilling and bending aluminum. You should always clamp metal pieces tightly in the vise when working on them with other tools. It is dangerous to try drilling metal pieces that are not clamped in a vise.

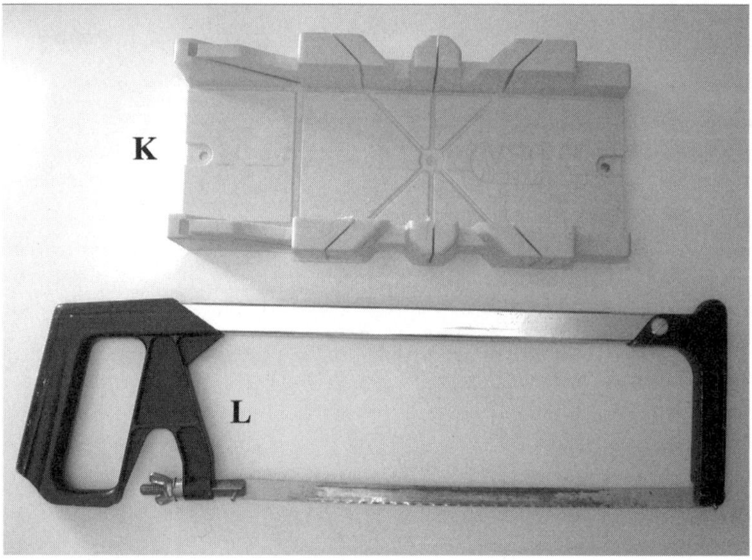

FIGURE 3.1 Hacksaw and miter box.

FIGURE 3.2 Band saw fitted with a metal cutting blade.

FIGURE 3.3 Work bench vise.

An electric drill will be needed during the mechanical construction phase of building the robot and the fabrication of the printed circuit boards. You will be required to drill approximately 150 holes during the process of creating your robot. An electric hand drill like the one shown in Figure 3.4 can be used.

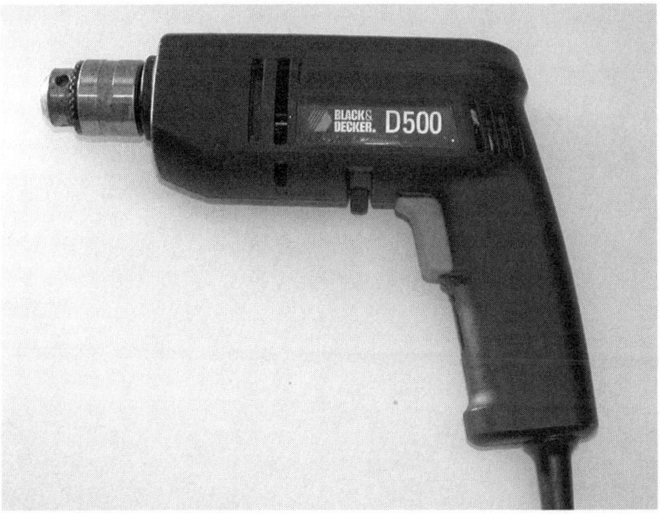

FIGURE 3.4 A handheld electric drill.

If you do plan on building robots as a hobby, then a small drill press like the one shown in Figure 3.5 would be a great idea. It is highly recommended that a drill press be used when drilling holes in printed circuit boards where accuracy and straightness are important. These small drill presses don't cost much more than a good electric hand drill. I have added an adjustable X-Y vise to the drill press in my workshop. This makes it possible to mill aluminum if an endmill, like the one shown in Figure 3.6, is purchased from a machine shop supplier. The drill press can then double as a small milling machine.

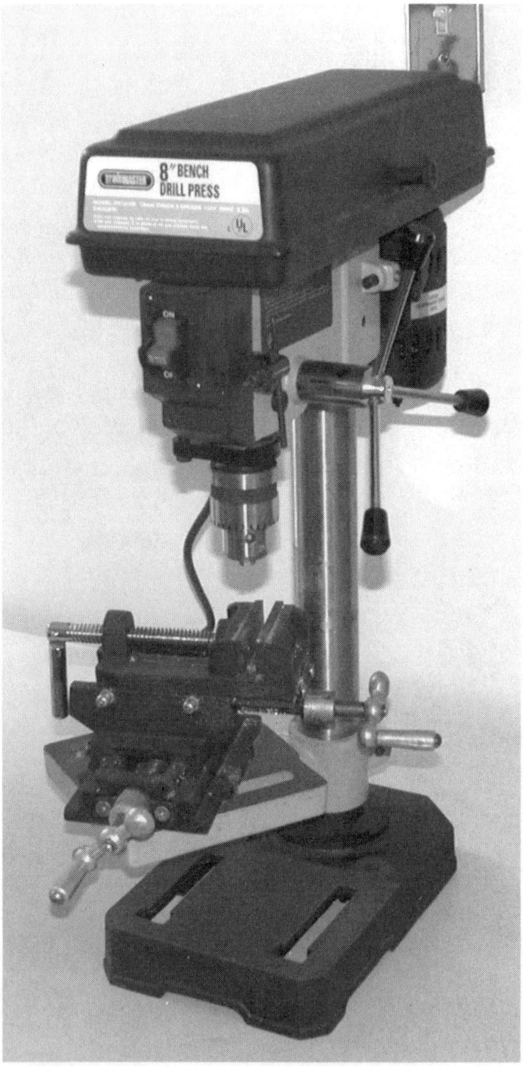

FIGURE 3.5 A small electric drill press with an X-Y adjustable vise.

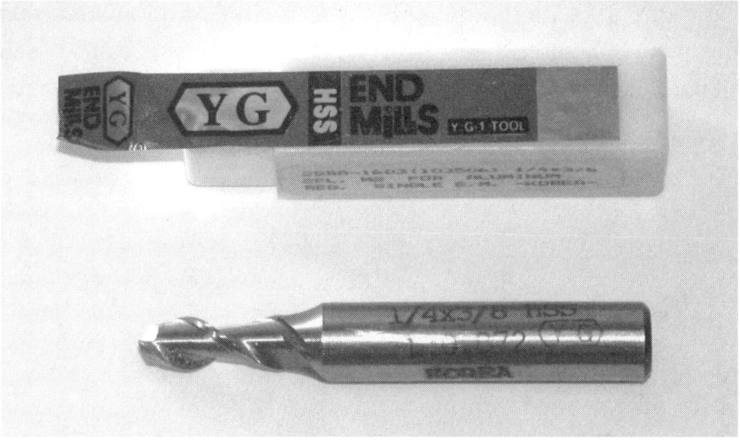

FIGURE 3.6 Aluminum-cutting endmill.

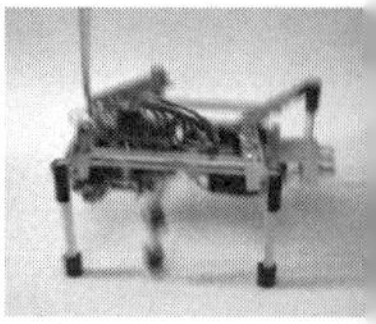

You will need a set of drill bits like the ones pictured in Figure 3.7. The two drill bits used most often during this project are the 5/32-inch and the 1/4-inch bits. You will need to separately buy the small 1/32-inch and 3/64-inch bits that will be used to drill the component holes in the printed circuit boards.

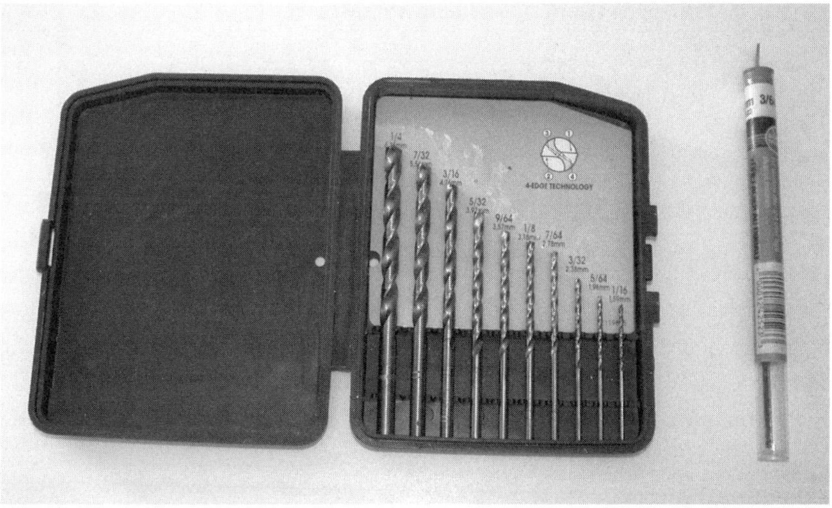

FIGURE 3.7 Drill bit set.

An adjustable wrench (marked E in Figure 3.8), a pair of side cutters (F), a pair of pliers (G), needle nose pliers (H), a Phillips screwdriver (I), and a Robertson screwdriver (J) will all be needed at some point during the construction of the robot. A set of miniature screwdrivers might be useful as well. The needle nose pliers can be used to hold wire and small components in place while soldering, bending wire, and holding machine screw nuts.

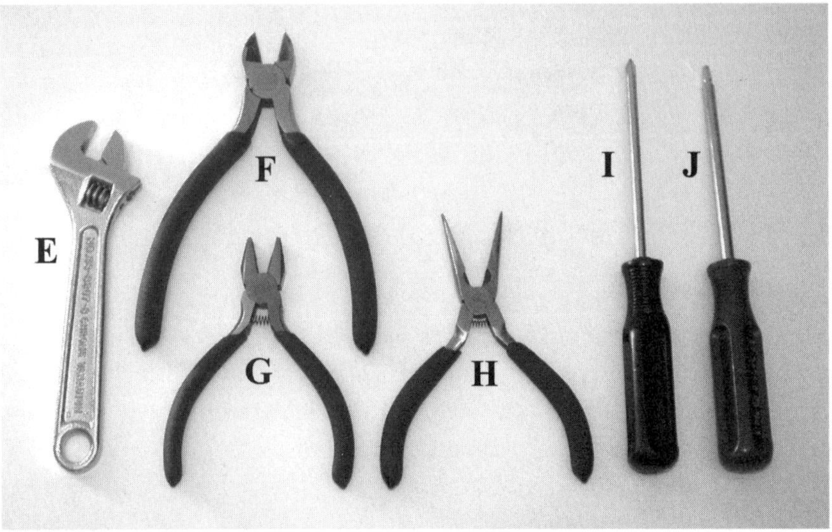

FIGURE 3.8 Various pliers, a wrench, and screwdrivers.

The wire strippers shown in Figure 3.9 (A) are used to strip the protective insulation off of wire without cutting the wire itself. The device is designed to accommodate a number of wire sizes. A pair of wire cutters (C) will be needed to cut wire when fabricating jumper wires and wiring power to the circuits. Rosin-core solder (B) will be needed when soldering components to the circuit boards, creating jumper wires, and wiring the battery connectors and power switch. To make soldering components to the printed circuit boards as easy as possible, buy the thinnest solder that you can find. A chip-pulling tool (D) will definitely be needed for removing the PIC 16F84 chip from the 18-pin socket. The PIC will be inserted and removed from the socket on the main controller board many times, as the software is changed and the PIC reprogrammed during experiments.

An adjustable work stand like the one shown in Figure 3.10 (M) will be useful when soldering components to circuit boards, or holding wires when soldering header connectors to the bare wires. A utility knife (N) will also be helpful when cutting heat-shrink tubing or small parts.

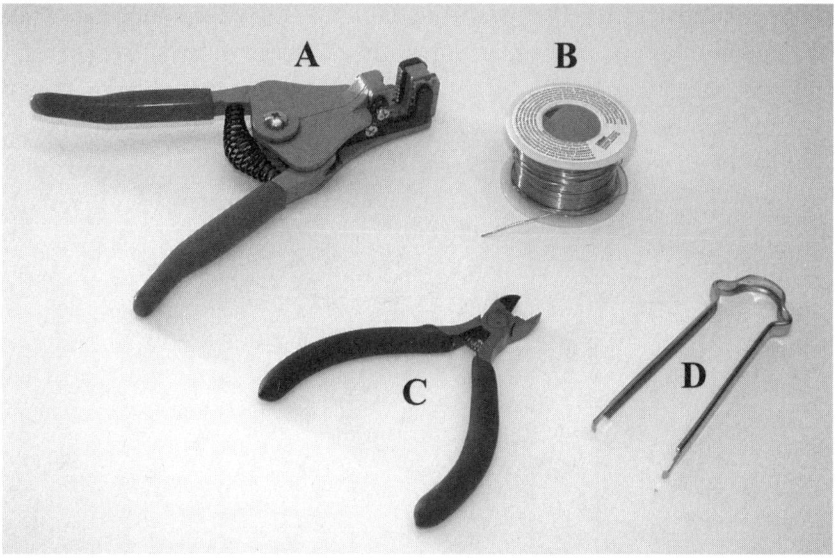

FIGURE 3.9 Wire strippers, cutters, solder, and a chip-pulling device.

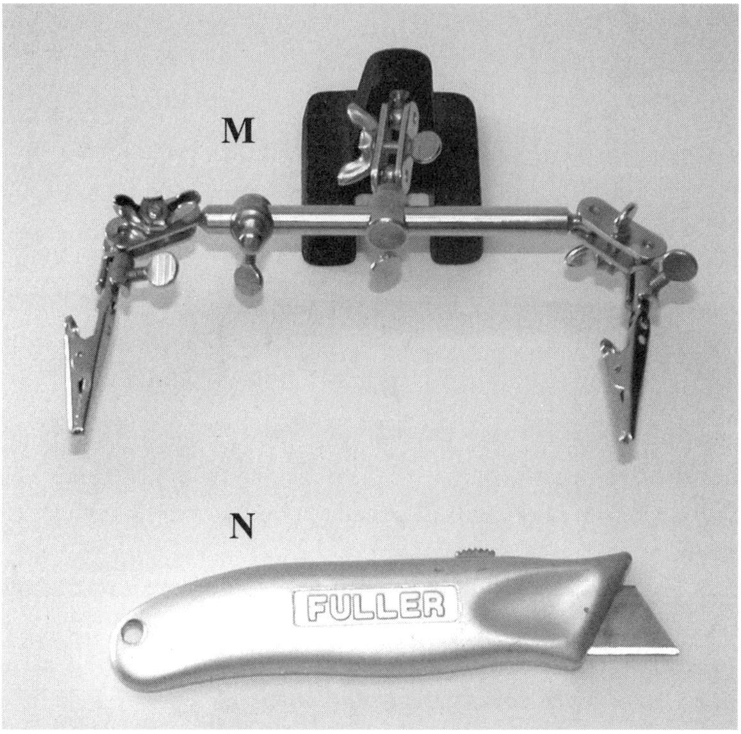

FIGURE 3.10 Adjustable work stand and utility knife.

A soldering iron similar to the one shown in Figure 3.11 will be required when building the main controller circuit board and the infrared sensor board. An expensive soldering iron is not necessary, but the advantage to buying a good soldering iron is that the temperature can be set. A 15–25-watt pencil-style soldering iron will do the job and will help to protect delicate components from burning out.

FIGURE 3.11 **Soldering iron with adjustable temperature.**

An adjustable square (O) and a good ruler (P) will be required when measuring the cutting and drilling marks on the aluminum pieces that make up the robot's body and legs. A hot glue gun (Q) and glue sticks will be needed at certain points in the construction. See Figure 3.12.

A hammer (shown in Figure 3.13; R) will be needed for bending aluminum along with a metal file (S) to smooth the edges of metal pieces after they have been cut or drilled. A tube of quick-setting epoxy (T) may be used to secure parts. Safety glasses (U) should be worn at all times when cutting and drilling metal or soldering.

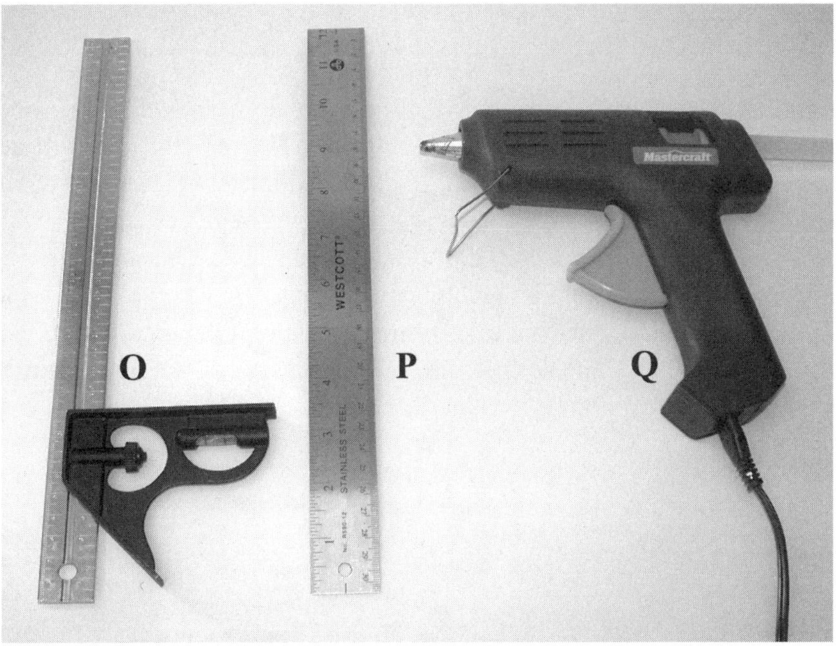

FIGURE 3.12 Adjustable square, ruler, and glue gun.

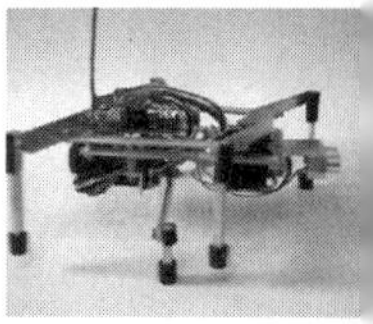

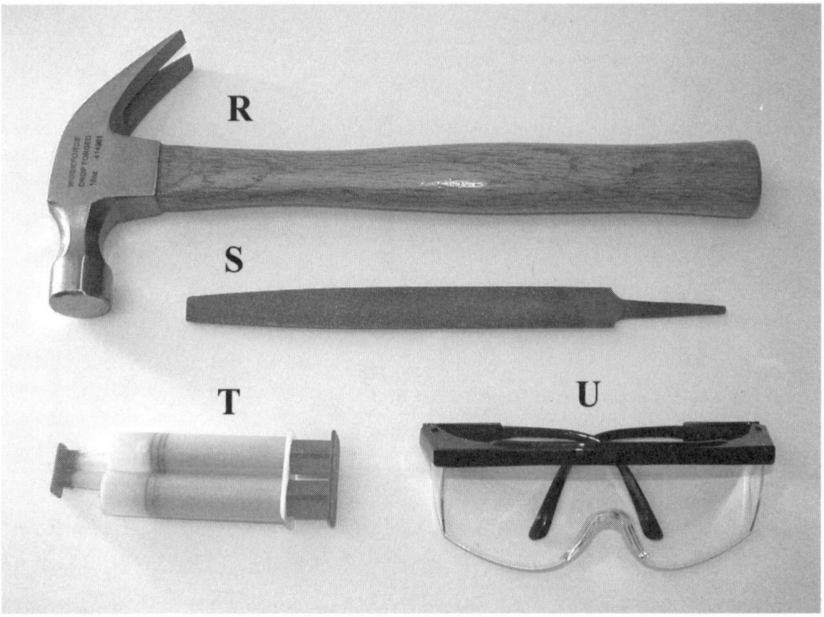

FIGURE 3.13 Hammer, file, epoxy, and safety glasses.

Test Equipment

To calibrate and troubleshoot the electronics, a digital multimeter with frequency counting capabilities similar to the Fluke 87 multimeter (Figure 3.14, left) will be needed. When working with electronic circuits, a good multimeter is invaluable. The second multimeter in Figure 3.14 (right) is manufactured by Circuit Test and it measures capacitance, resistance, and inductance. It is nice to be able to measure the exact values of components when working on precise circuits, but in most cases this is not necessary. If you are winding your own transformers or chokes, the ability to measure inductance will be helpful. The specific use of the multimeter will be explained in Chapter 7.

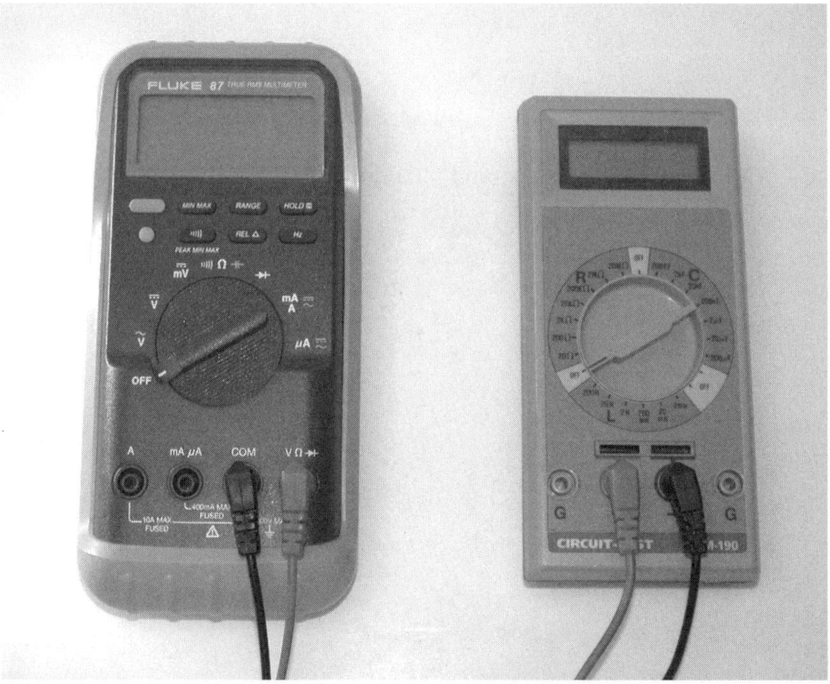

FIGURE 3.14 **Fluke and Circuit Test multimeters.**

If you are really serious about electronics, then an oscilloscope like the one pictured in Figure 3.15 is a great investment. This is the Tektronix TDS 210 dual channel, digital real-time oscilloscope with a 60-MHz bandwidth. The TDS 210 on my bench also has the RS-232, GPIB, and centronics port module added so that a hard copy of waveforms can be output. The great advantage to using an oscilloscope is the ability to visualize what is happening with a circuit. The new digital oscilloscopes also automatically calculate things like the frequency, period, mean, peak to peak, and true RMS of a waveform. Two other pieces of equipment that you will probably need to use quite often are a regulated direct current power supply and a function generator.

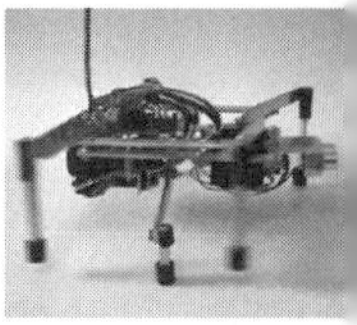

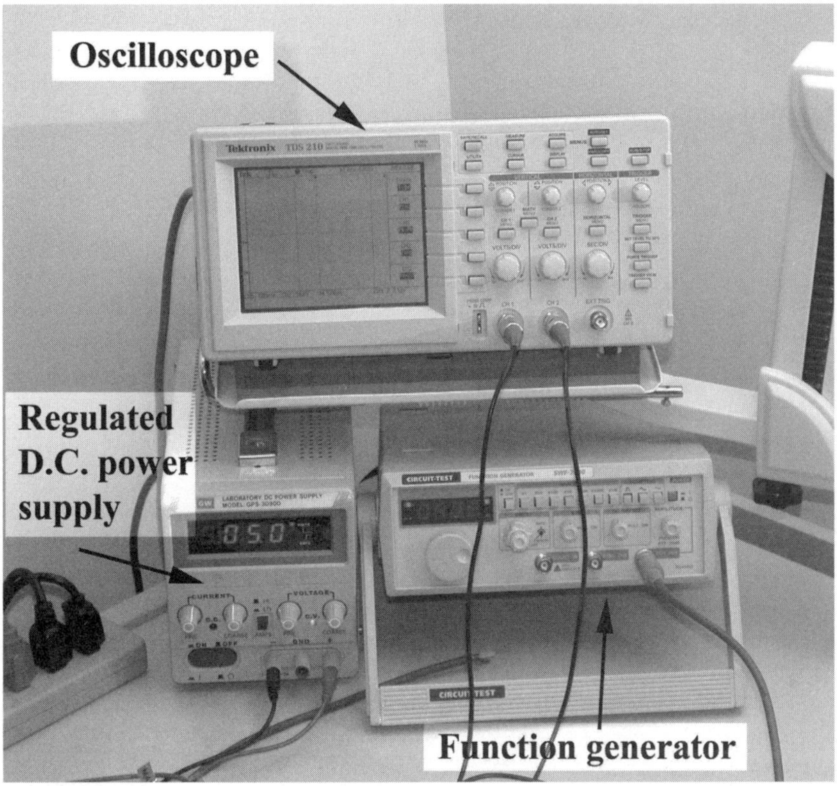

FIGURE 3.15 Oscilloscope, regulated D.C. power supply, and a function generator.

None of the equipment shown in Figure 3.15 is absolutely required when building the robot in this book, but this equipment will make your life as an electronics experimenter much easier. There is nothing more frustrating than finding out that a circuit you are working on is malfunctioning because of a dead battery or an oscillator calibrated to the wrong frequency. If a good power supply and oscilloscope are used when building and testing a circuit, the chance of these kinds of problems surfacing is much lower. I have always found that if I am working late at night and start to encounter a lot of small problems and make mistakes, the best thing to do is to shut my equipment down and get a good night's sleep. Sometimes the difference between frying an expensive chip or the circuit's working perfectly on the first try is just one misplaced component.

Construction Materials

The robot's body is constructed using aluminum and fasteners that are readily available at most hardware stores. There are five sizes of aluminum that will be used. The first stock measures 1/2 inch wide by 1/8 inch thick, and is usually bought in lengths of 4 feet or longer. Most of the robot's body is constructed from aluminum with the dimensions as shown in Figure 3.16.

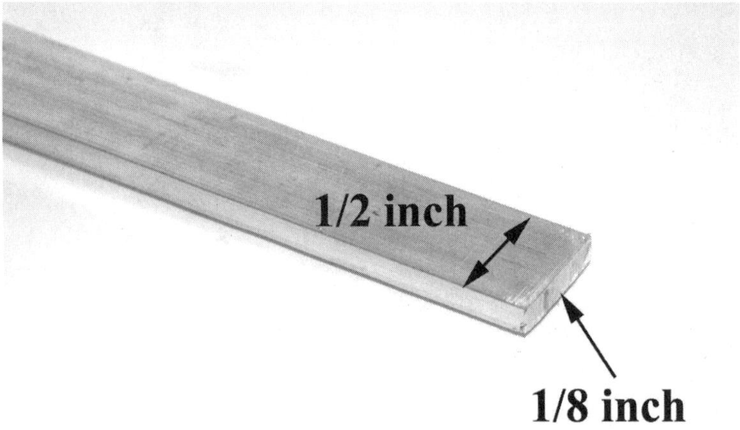

FIGURE 3.16 1/2 inch by 1/8 inch aluminum stock.

The second type of aluminum stock that will be used measures 1/4 inch × 1/4 inch, and is shown in Figure 3.17. It is usually bought in lengths of 4 feet or longer as well.

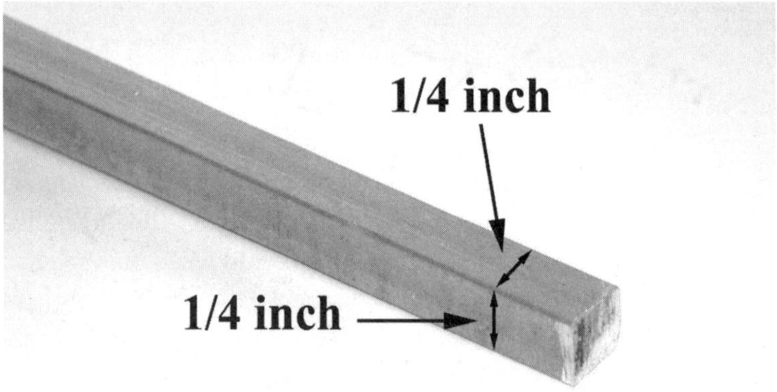

FIGURE 3.17 Aluminum stock with 1/4 inch by 1/4 inch dimensions.

The third kind of aluminum stock is 1/2 inch × 1/2 inch angle aluminum and is 1/16 inch thick as shown in Figure 3.18. Part of the robot's legs will be made using this stock.

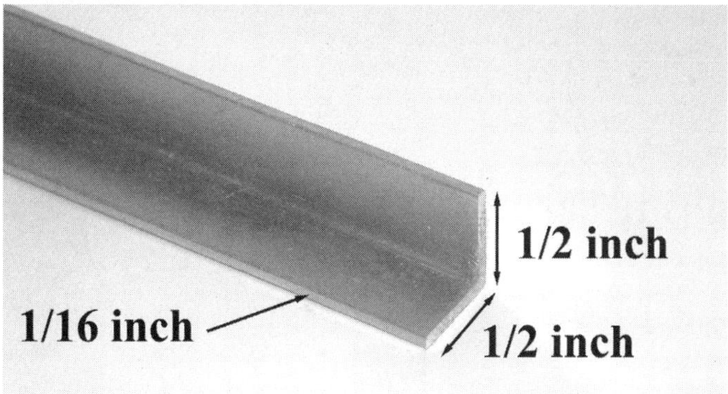

FIGURE 3.18 1/2 inch angle aluminum.

The fourth type is 1/16 inch thick flat aluminum, as shown in Figure 3.19, and it is usually bought in larger sheets. The fifth type of stock that will be needed is 3/4 inch × 3/4 inch angle aluminum. However, most metal suppliers will cut it down for you. This thickness of aluminum is great for cutting out custom parts and it is easy to bend, making it ideal for the hobbyist experimenter. I buy all of my metal from a Canadian company called *The Metal Supermarket* (www.metalsupermarkets.com) because their prices are much lower than if you buy metal at a hardware store. Their friendly staff is always helpful, and will cut the stock to whatever size you require. I usually get them to cut the raw stock in half so that it will fit into the back seat of my car.

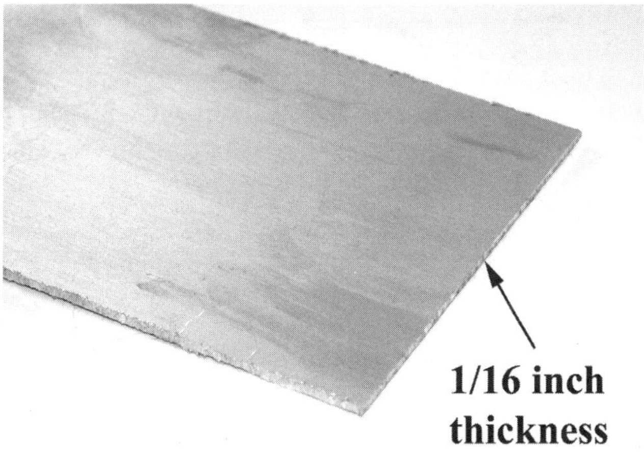

FIGURE 3.19 1/16 inch thick flat aluminum.

The fasteners that we will be using are 6/32-inch diameter machine screws, nuts, lock washers, locking nuts, and nylon washers as shown in Figure 3.20. Three different lengths of machine screws will be used: 1 inch, 3/4 inch, and 1/2 inch.

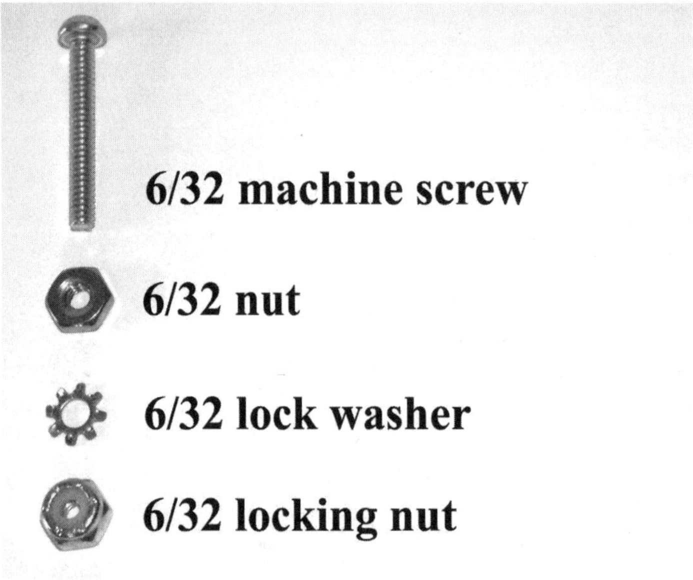

FIGURE 3.20 6/32-inch diameter machine screw, lock washer, nuts, and nylon washer.

Summary

Now that all the tools, test equipment, and materials necessary to build robots have been covered, it is time to start into the construction of an actual autonomous walking robot. In the next chapter, the stages of the robot's evolution will be outlined so that you can get an idea of how the construction will proceed.

4

STAGES OF ROBOTIC EVOLUTION

The six-legged robot detailed in this book will be built in evolutionary stages as construction progresses. Each development can be thought of as an evolutionary step in the process of creating an insect-like artificial lifeform. When a certain amount of work has been done, a number of experiments will be performed with the robot in that configuration. As the robot advances, it will be capable of dealing with and reacting to sensory input. It will also be able to map its environment and respond accordingly.

The ten stages are:

1. **Mechanical construction.** This is the genesis stage where the robot's body is constructed. The chassis that is built will carry the servos, legs, electronics, sensors, and a microprocessor that will make up the robot. Figure 4.1 shows the completed chassis with servos and legs.
2. **The main controller circuit board.** This circuit board contains all the support circuitry for the PICmicro 16F84 MCU, such as a regulated 5V power supply, a piezo speaker, light-emitting diodes, and input/output connectors for the leg servos and sensors. This can be thought of as the robot's brain. The controller board is shown in Figure 4.2.

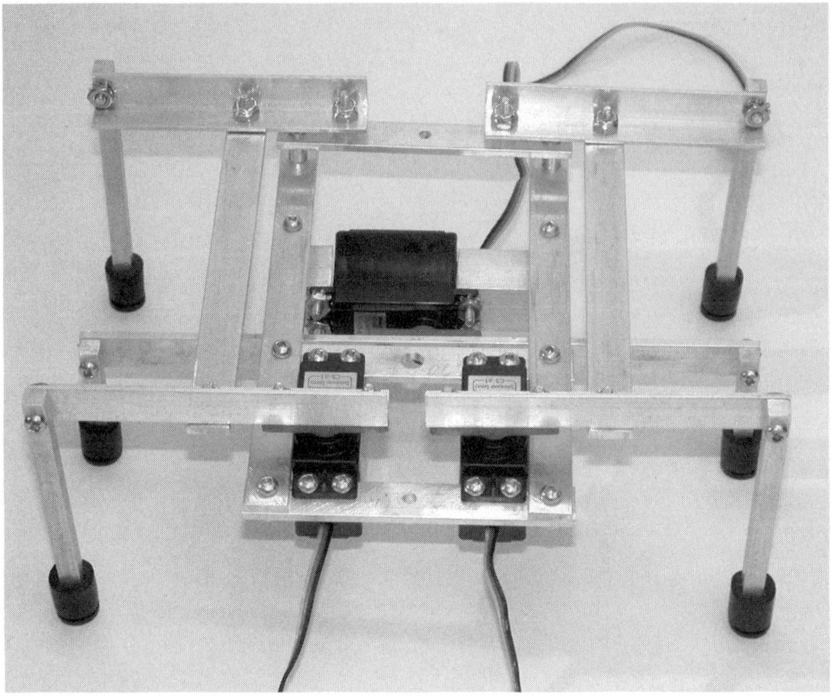

FIGURE 4.1 Completed robot chassis.

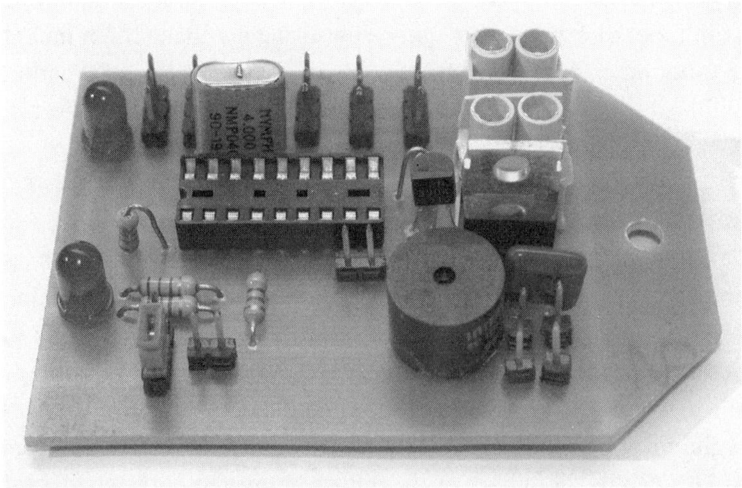

FIGURE 4.2 Main controller circuit board.

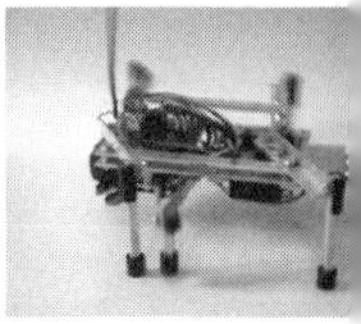

3. The Infrared sensor board. This board will give the robot an electronic sense of sight for obstacle-avoidance capabilities necessary for survival (Figure 4.3).

FIGURE 4.3 **The infrared sensor board gives the robot a sense of sight.**

4. Integration. At this stage in the robot's evolution, all the individual components are integrated and can communicate (Figure 4.4). This can be thought of as the robot's "nervous system."

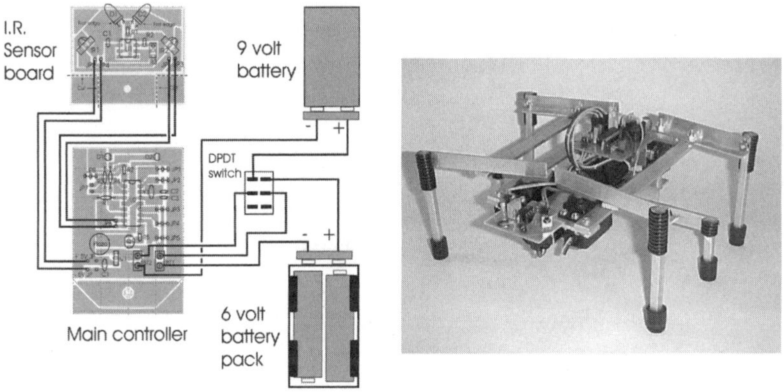

FIGURE 4.4 **The electronic and mechanical systems integrated.**

5. **Programming.** At this stage, microcontroller programming is covered, giving the robot the ability to think and control its actions. Figure 4.5 shows a Microchip PIC 16F84 microcontroller that will function as the robot's brain.

FIGURE 4.5 The PIC 16F84 microcontroller brain of the robot.

6. **Evolution.** The robot learns to walk, avoid obstacles, explore the environment, and unique behaviors develop. Figure 4.6 shows the robot avoiding obstacles while walking.

FIGURE 4.6 Robot walking while exploring its surroundings.

7. **Ultrasonic detection.** At this stage, the robot uses sonar much as a bat or dolphin does to create maps of the surrounding areas. The robot now stores an internal representation of the outside world in its memory, which is considered by some to be the first measure of intelligence. The robot is able to use this internal representation of the world to map the areas around itself and make intelligent decisions about navigation. The robot is then programmed to solve simple mazes. Figure 4.7 shows the ultrasonic ranger that will give the robot room-mapping capabilities and much better navigational skills.

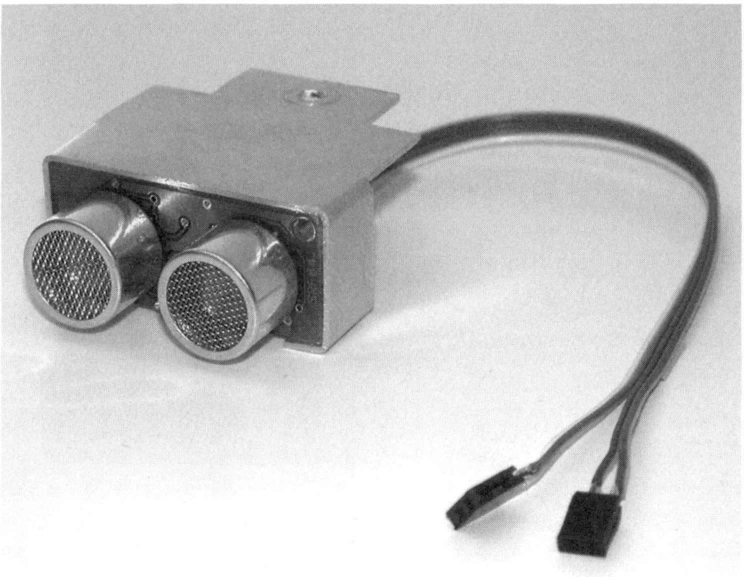

FIGURE 4.7 Sonar ranging device.

8. **Radio remote control.** When the radio receiver is added, the robot (Figure 4.8) will have the capability to be controlled remotely up to a distance of one mile. The robot can be sent out to remote locations to perform tasks at the will of the operator. Programs will be written to interpret the radio control signals and have the robot perform the appropriate actions. A certain amount of autonomy can still be built into the actions of the robot after it receives a radio control command.
9. **Robotic gripper.** In the final stage of the robot's evolution, a robotic gripper is added (Figure 4.9) so that it can pick up objects and move them around. The gripper itself is under complete remote control by the human operator. The walking gait sequences are still controlled by the microcontroller, but the operator determines the movement commands.

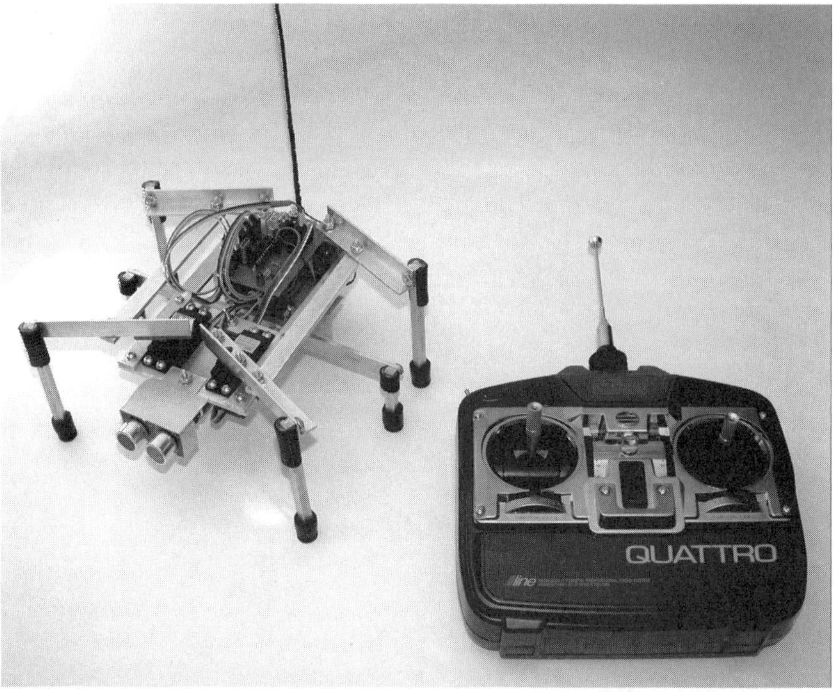

FIGURE 4.8 Robot with remote control transmitter and receiver integrated.

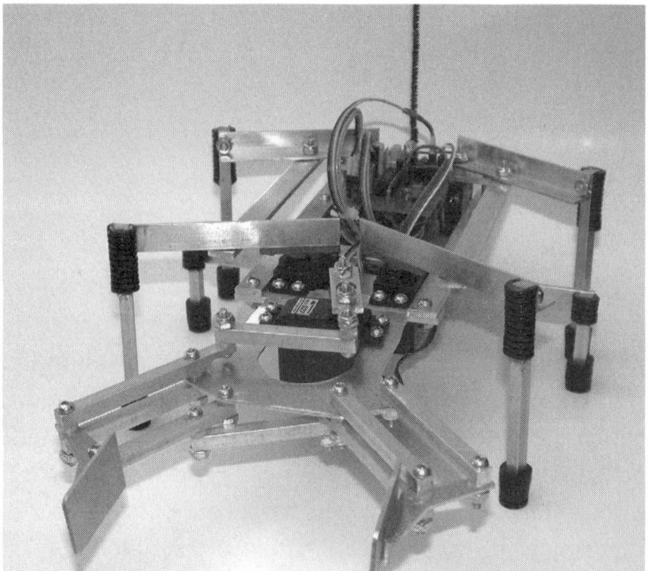

FIGURE 4.9 Robot with gripper enhancement.

10. **Customization.** This stage includes ideas to improve and customize the robot. This phase of the robot's evolution is up to you, and is limited only by your imagination. In Chapter 14, there are many suggestions for enhancements and modifications. Figure 4.10 shows a group of robots working together as a swarm.

FIGURE 4.10 A swarm of Insectronics.

Summary

With an overview of the robot's evolution behind us, the mechanical construction of the robot will begin. In Chapter 5, the robotic chassis will be constructed using some of the tools and materials that were discussed in Chapter 3. The robot chassis will support the servos, legs, electronics, sensors, and eventually a gripper.

5

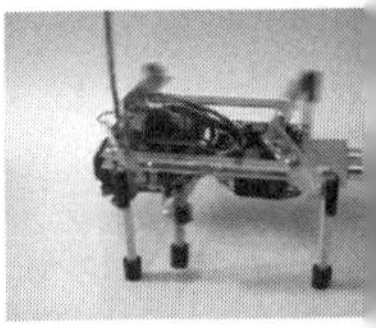

CHASSIS, LEGS, AND SERVO MOUNTS

Constructing the Chassis

This chapter deals with the construction of the robot's aluminum chassis, legs, servo mounts, and linkages. This will require the use of hand tools such as a hacksaw, file, electric drill, screwdriver and pliers. It would also be a good idea to have access to a workbench with a vise when cutting and drilling the aluminum pieces. Table 5.1 offers a detailed list of all the parts you'll need.

First, cut and drill the five pieces of aluminum that will make up the body in which three servos will eventually be mounted. Figure 5.1 shows the cutting measurements for each of the five aluminum pieces that make up the Insectronic's body. Using a piece of the 1/2 × 1/8-inch aluminum, cut the strips with a hacksaw and use a miter box to achieve straight cuts. Alternately, you can use a band saw with a metal cutting blade to save time and get more accurate and clean cuts. Each piece should be marked with the letter that is beside it in Figure 5.1 to identify its position when the individual parts are assembled later (A,B,C,D,E). Use a file to smooth the rough edges after cutting. Figure 5.2 shows the cut and drilled pieces.

TABLE 5.1 Parts List for Chassis Construction

PARTS	QUANTITY
1 × 1/8-inch aluminum	22.5 inches
~~1/4-diameter aluminum tubing~~	~~1 inch~~
1/4 × 1/2-inch standoffs	2
6/32 × 1/2-inch machine screws	12
6/32 × 3/4-inch machine screws	2
6/32-inch nuts	14
6/32-lock washers	14
Standard Hobbico servo CS-61	2

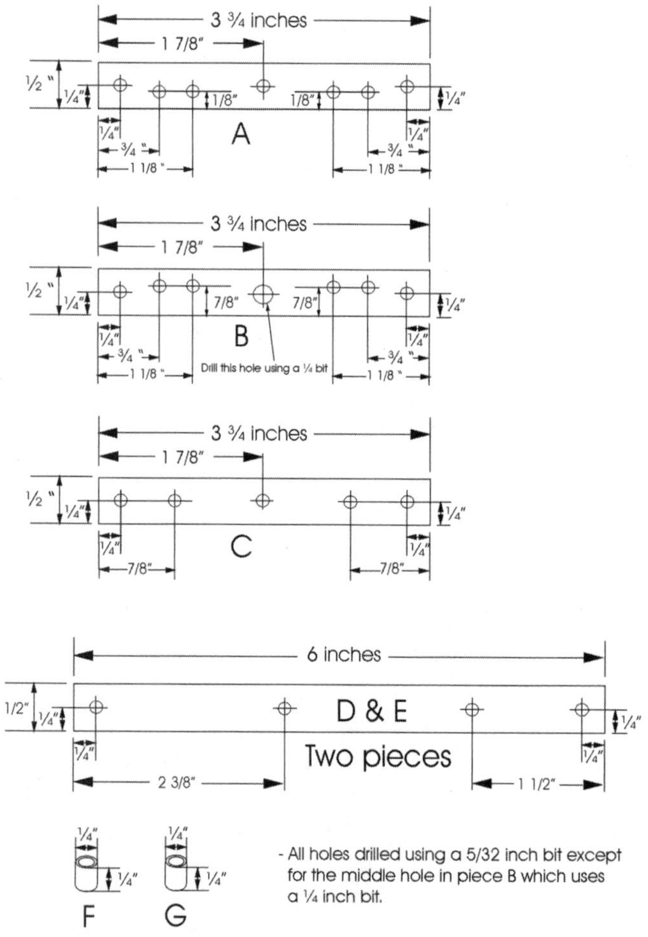

FIGURE 5.1 Cutting and drilling guide for the robot chassis.

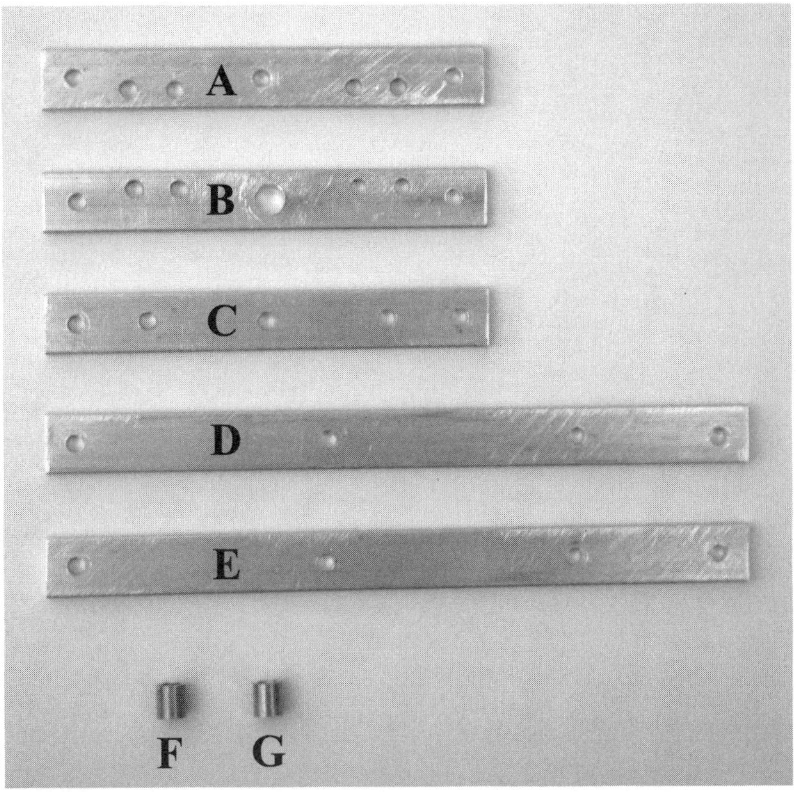

FIGURE 5.2 Cut and drilled aluminum for the body chassis.

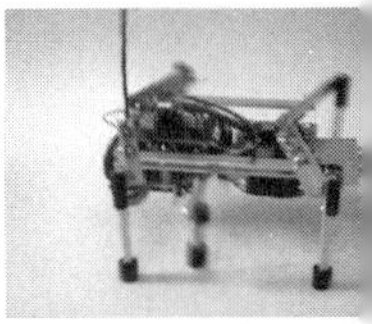

When each of the five pieces has been cut, use a ruler to accurately measure and mark the positions where the holes will be drilled. All holes should be drilled using a 5/32-inch drill bit, except for the middle hole in part B, which is drilled with a 1/4-inch bit. If you wish to paint the aluminum, now would be the best time to do so.

Using the 1/4-inch diameter hollow aluminum tubing, cut two pieces 1/4 inch in length (identified as pieces F and G in Figure 5.1).

To assemble the individual pieces, place part A and part B on a flat surface and lay pieces D and E on top, aligning the holes. See Figures 5.3 and 5.4 as a guide. Fasten the pieces together using the 6/32 × 1/2-inch machine screws, lock washers, and nuts as illustrated in Figure 5.3. Take the two standoffs (parts F and G) and place them over the two holes at the ends of pieces D and E. Place piece C on top of the two standoffs and secure it using the 6/32 × 3/4-inch machine screws, nuts, and lock washers as shown in Figures 5.2, 5.4, and 5.5. Place two of the hobby servos in place between pieces A and B (as shown in Figure 5.3), line up the dill holes, and secure with 6/32 × 1/2-machine screws, lock washers, and nuts. Now that the chassis is completed, the next step is to construct the middle leg servo mounts.

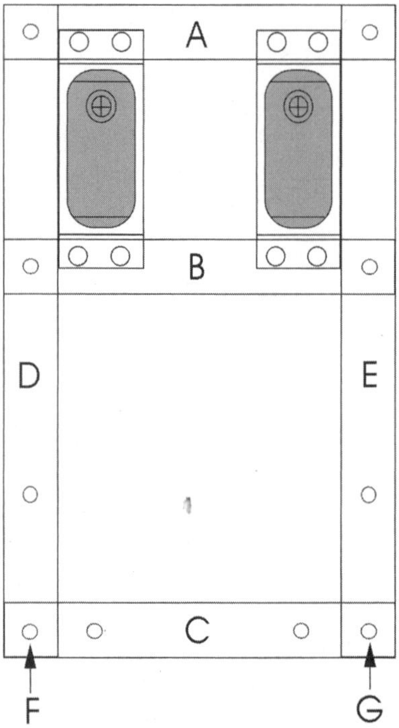

FIGURE 5.3 Parts placement for the body chassis.

FIGURE 5.4 Assembled chassis.

FIGURE 5.5 Machine screws with lock washers and nuts.

Constructing the Middle Leg Servo Mounts

Table 5.2 provides an itemized list of the parts you'll be using to construct the leg servo mounts.

TABLE 5.2 Parts List for Middle Leg Servo Mounts

PARTS	QUANTITY
3/4-inch angle aluminum	2 inches
1/4-inch diameter aluminum tubing	1 inch
6/32 × 1/2-inch machine screws	4
6/32 × 1-inch machine screws	2
6/32-inch nuts	6
6/32-lock washers	6
1/2-inch black wire protector	12 inches
Standard Hobbico servo CS-61	1

The next subassembly is the servo mounts for the middle leg. Using a piece of the 3/4 × 3/4-inch angle aluminum, cut two pieces as shown in Figure 5.6. The first piece will be labeled part H, and is 1 1/4 inches long. The second piece is cut to a length of 3/4 inch, and is labeled part I. Use the drilling guide in Figure 5.6 to accurately measure, mark, and drill the required holes. All holes are drilled using a 5/32-inch drill bit.

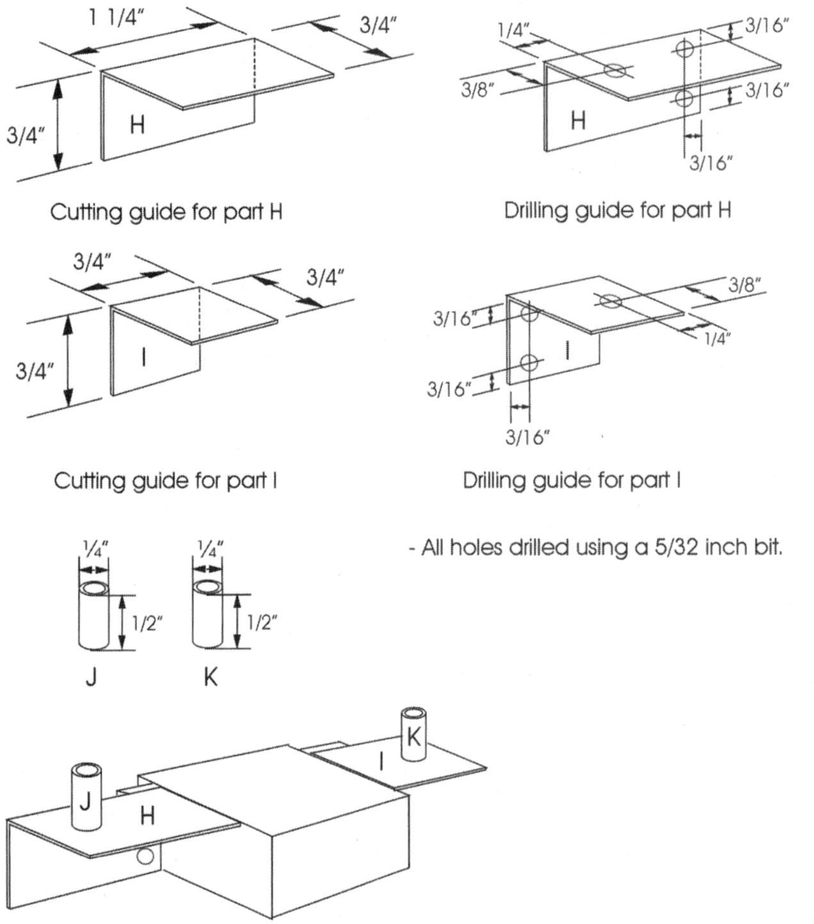

FIGURE 5.6 **Cutting and drilling guide for middle leg servo mounts.**

Figure 5.7 shows the completed parts cut and drilled.

Attach the servo mounts to the servo as shown in Figures 5.6, 5.8, and 5.9 using four of the 6/32 × 1/2-inch machine screws, lock washers, and nuts. Note that piece H is attached to the side of the servo closest to the servo shaft, as shown in Figure 5.8. Using the 1/4-inch diameter hollow aluminum tubing, cut two pieces 1/2 inch in

FIGURE 5.7 Cut and drilled middle leg servo mounts.

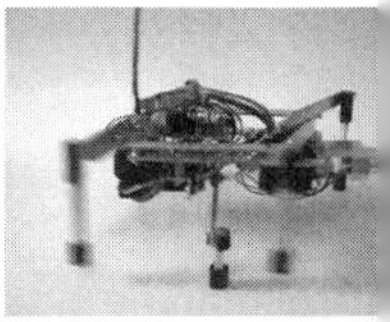

length (identified as pieces J and K). Line pieces J and K up with the holes on pieces H and I, and then attach to the chassis with two 1-inch machine screws going through the mounting holes on the chassis, the standoffs (pieces J and K), and through the holes on pieces H and I as shown in Figures 5.9 and 5.10. Secure with two 6/32 lock washers and nuts. When this stage is complete, you should have assembled the chassis as pictured in Figure 5.9.

FIGURE 5.8 Middle leg servo mounts attached to servo.

FIGURE 5.9 Middle leg servo mounts with servo and standoffs attached to chassis.

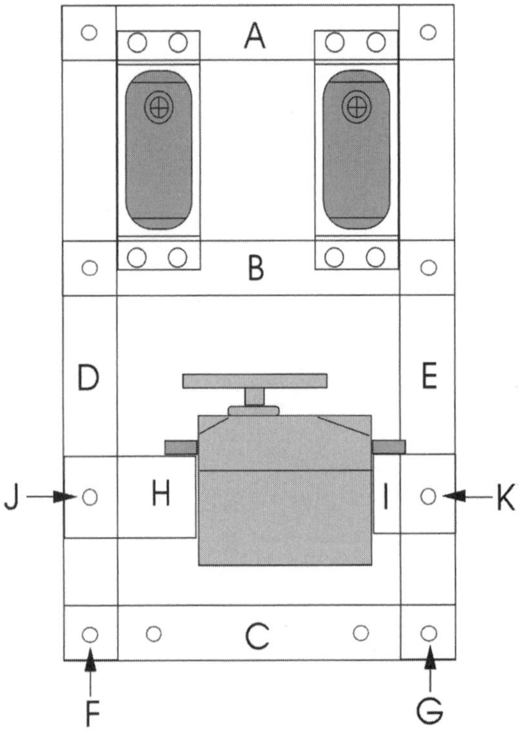

FIGURE 5.10 Parts placement for chassis with middle servo.

Construction of the Legs

Table 5.3 shows exactly what is required to add the legs.

TABLE 5.3 Parts List for the Construction of Insectronics' Legs

PARTS	QUANTITY
1/2-inch angle aluminum	15 inches
1/2 × 1/8-inch aluminum	19 inches
1/4 × 1/4-inch aluminum	16 inches
6/32 nylon washers	6
6/32 × 1/2-inch machine screws	12
6/32 × 1-inch machine screws	2
6/32-inch locking nuts with nylon inserts	6
6/32-inch lock washers	8
6/32-inch nuts	8
Servo horns (supplied with the servos)	3
1/4-inch black wire protector	12 inches

Construction of the legs will be broken down into three sections. The first section involves modifying the servo horns to provide mounts for the legs to the servos. The second section involves cutting and drilling the aluminum for the legs and the third section deals with mounting the legs to the servos and chassis.

MODIFYING THE SERVO HORNS

Using Figure 5.11 as a guide, cut the four pieces off the servo horn using a hacksaw or a pair of side cutters (as shown in Figure 5.12) and then file the edges smooth. The two pieces that are to be cut off are marked with the numbers 2 and 4 in the plastic, and are slightly smaller in width than the other two. Drill two holes as indicated in Figure 5.12 using a 5/32-inch drill bit. The holes are drilled through the existing smaller holes; the third hole from the end of the horn on each side marked as 13 in the plastic. Complete this same procedure for all three servo horns. When finished, the cut and drilled servo mounts should look like those shown in Figure 5.14.

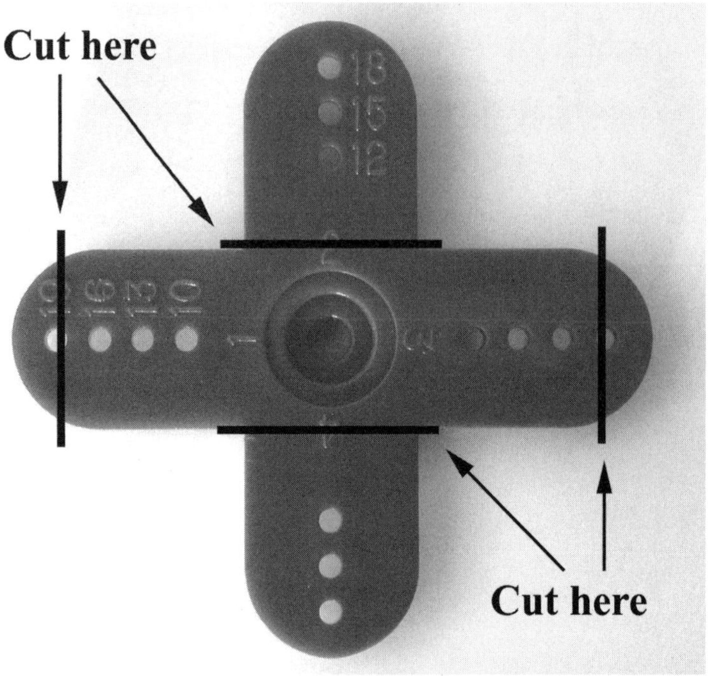

FIGURE 5.11 Servo horn modification cutting guide.

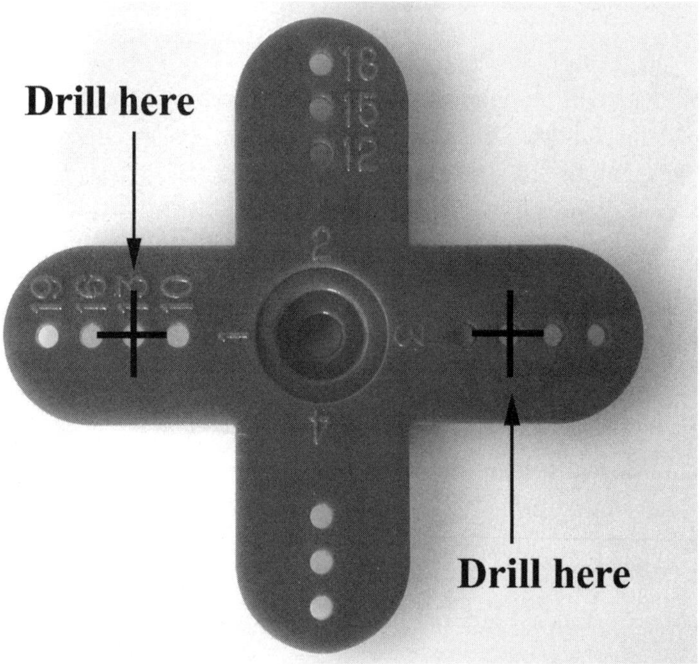

FIGURE 5.12 Servo horn modification drilling guide.

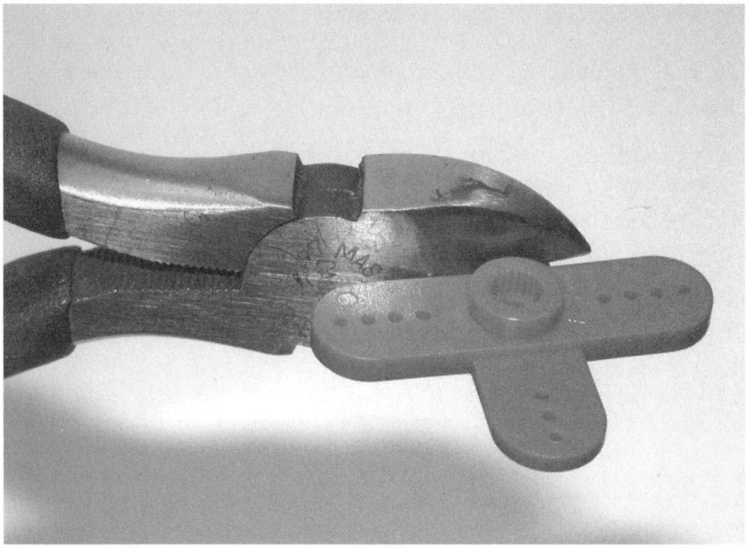

FIGURE 5.13 Cutting the servo horns with side cutters.

FIGURE 5.14 Completed servo mounts.

CONSTRUCTING THE LEGS

Using the 1/2-inch angle aluminum, cut and drill four pieces as indicated in Figure 5.15. Use a 5/32-inch drill bit. The four pieces shown in Figure 5.16 will be part of the four upper legs, and are marked as L, M, N, and O. Using the 1/2 × 1/8-inch aluminum, cut and drill two pieces 5 1/4 inches long as indicated in Figure 5.17. These two pieces will become the mechanical linkages that will control the hind legs. They are marked as P and Q. Using the 1/4 × 1/4-inch aluminum, cut and drill four pieces 3 1/2 inches long and two pieces 1 3/4 inches long as indicated in Figure 5.17. These are marked as R, S, T ,U, V, and W. These six pieces make up the lower legs and are shown along with pieces P and Q in Figure 5.18.

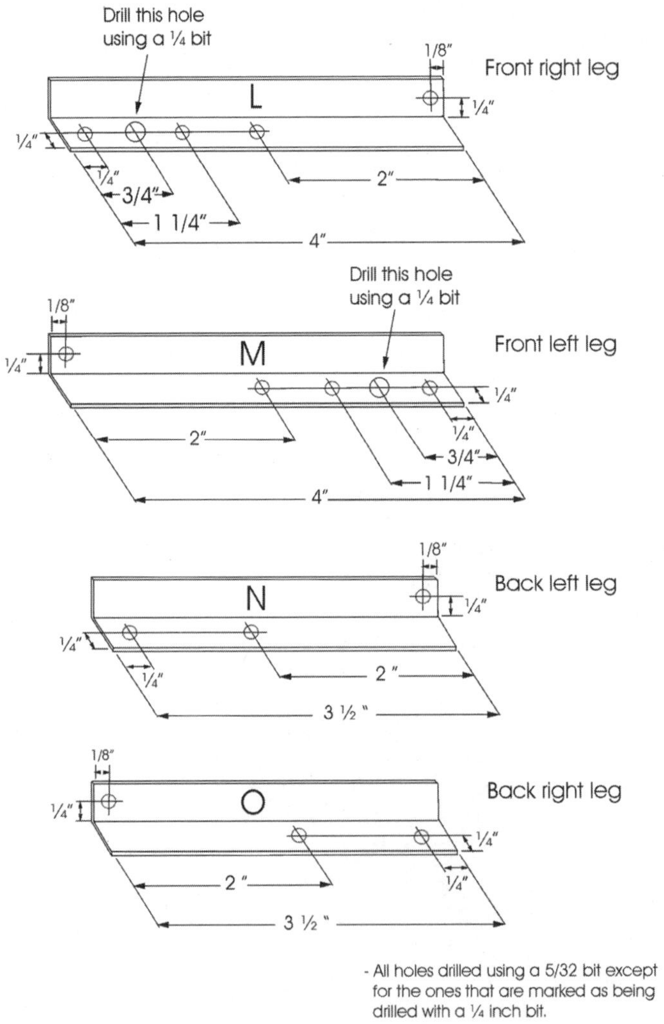

FIGURE 5.15 Upper leg cutting and drilling guide.

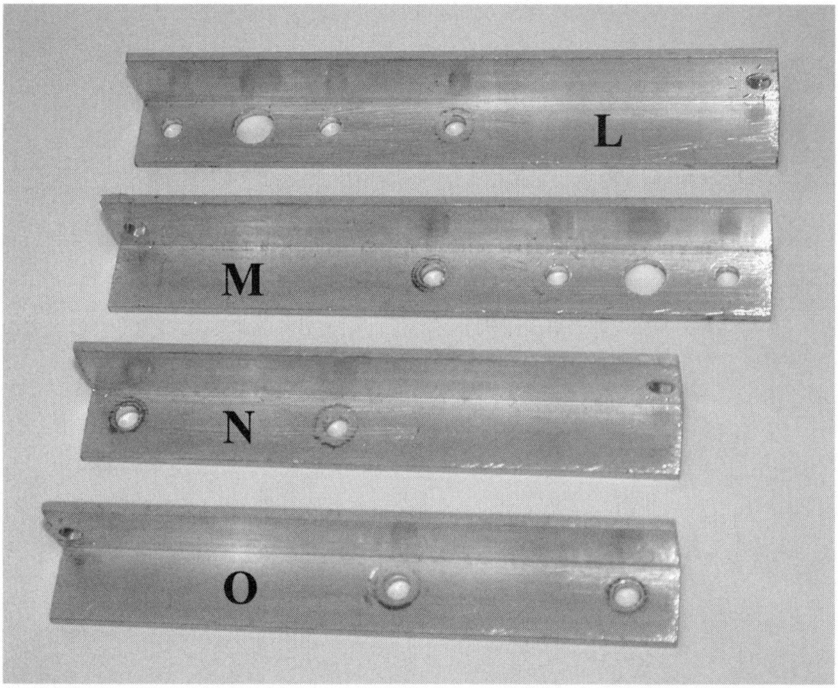

FIGURE 5.16 Cut and drilled upper leg pieces.

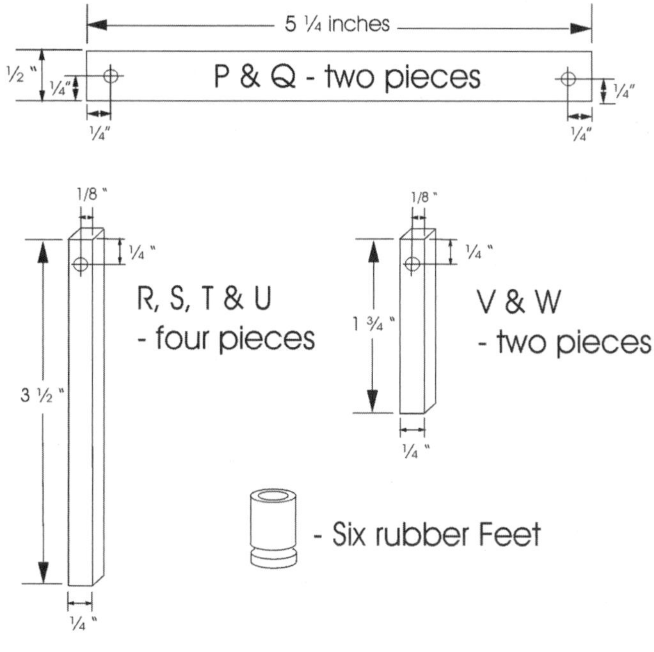

FIGURE 5.17 Mechanical linkages and lower leg cutting and drilling guide.

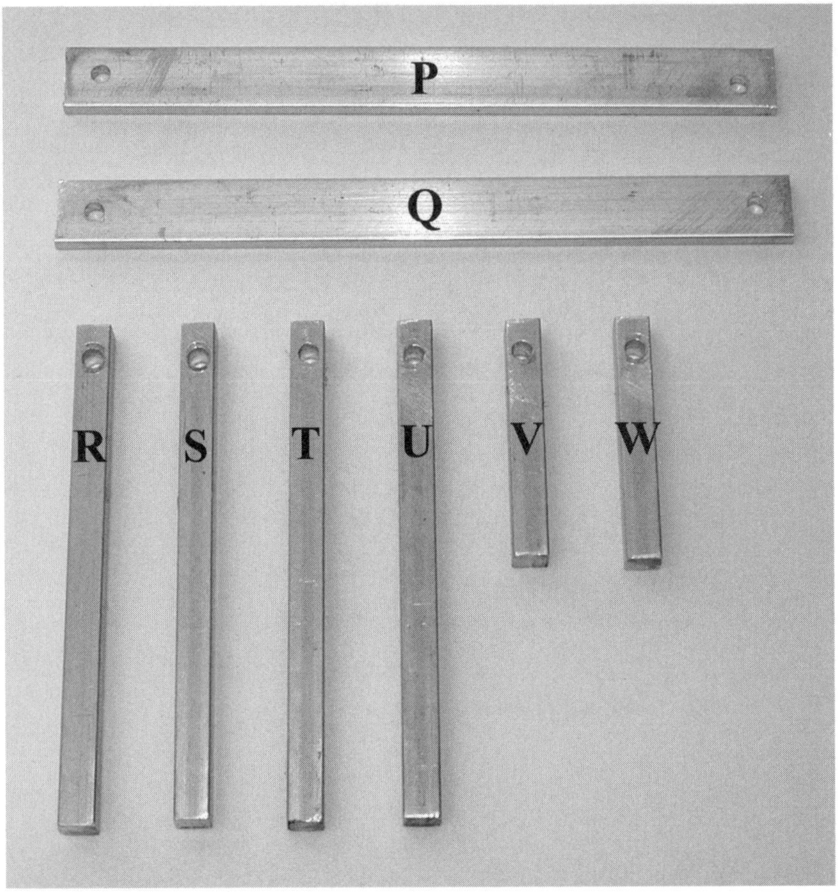

FIGURE 5.18 Cut and drilled linkages and lower leg pieces.

CONSTRUCTING THE MIDDLE LEG

Using a piece of the 1/2 by 1/8-inch aluminum, cut a piece 8 1/2 inches long and drill with a 5/32-inch bit as indicated in Figure 5.19. This piece is identified as X. This will be the pivoting middle section that the two middle legs will be attached to. The finished piece is shown in Figure 5.20.

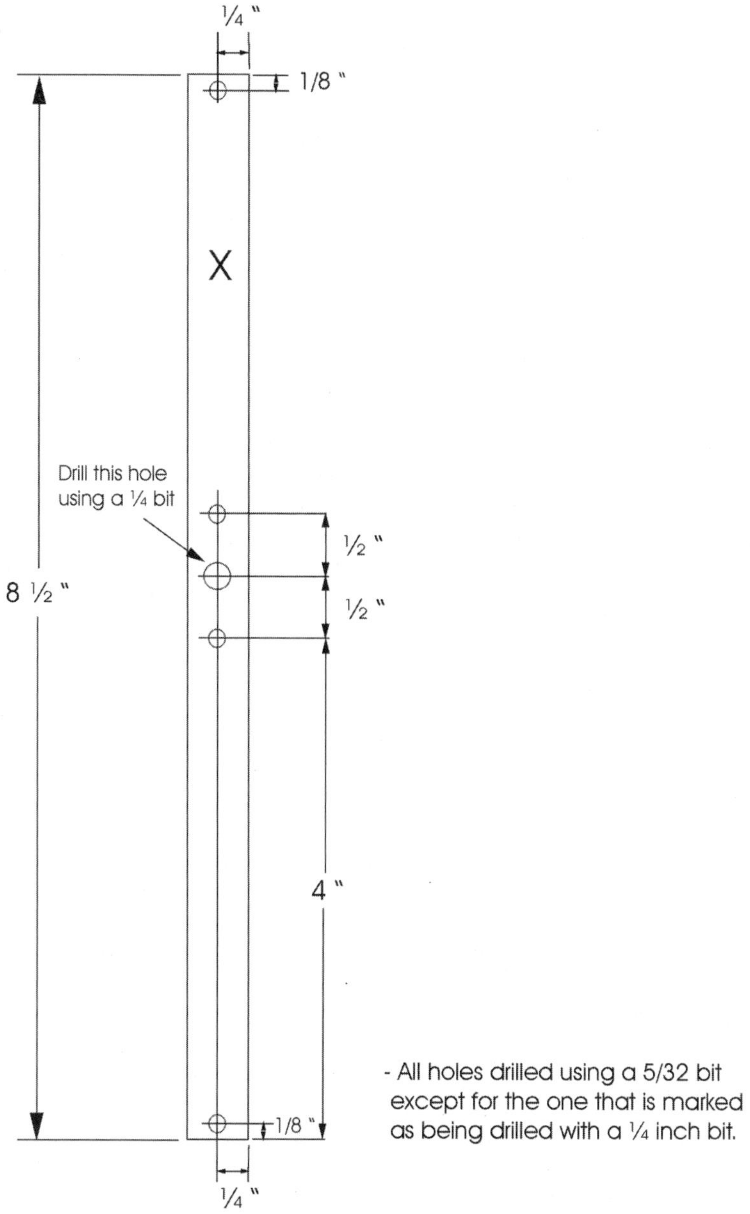

FIGURE 5.19 Middle leg cutting and drilling guide.

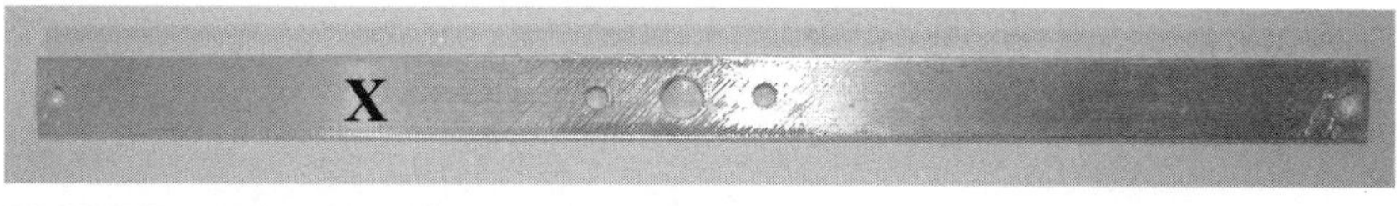

FIGURE 5.20 Middle leg cut and drilled.

Assembling Servo Horns and Leg Mechanics

Take upper leg pieces L and M and fasten a modified servo horn to each piece using two 6/32 × 1/2-inch machine screws, lock washers, and nuts for each as shown in Figure 5.21. Take the middle leg marked X (Figure 5.20) and fasten the third modified servo horn using two 6/32 × 1/2-inch machine screws, lock washers, and nuts as shown in Figure 5.22. Using Figure 5.23 as a guide, attach lower leg piece R to piece L with a 6/32 × 1/2-inch machine screw, lock washer, and nut. Do the same as the previous step with pieces S and M, using Figure 5.23 as a guide. Follow Figure 5.24 to attach lower leg piece T to piece N with a 6/32 × 1/2-inch machine screw, lock washer and nut. Do the same with pieces U and O, using Figure 5.24 as a guide. Figure 5.25 shows how to attach pieces V and W to middle leg piece X using two 6/32 × 1/2-inch machine screws, lock washers, and nuts.

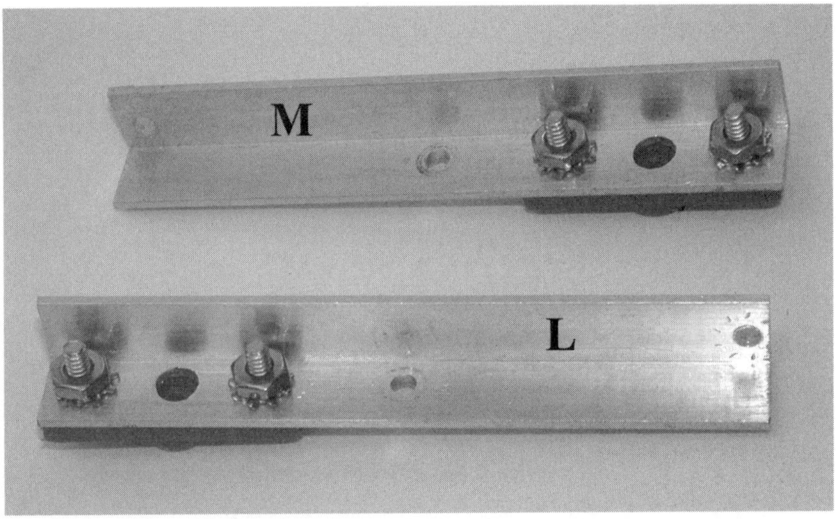

FIGURE 5.21 **Modified servo horns attached to upper leg pieces.**

FIGURE 5.22 Modified servo horn attached to middle leg cross piece.

FIGURE 5.23 Front upper and lower leg assembly guide.

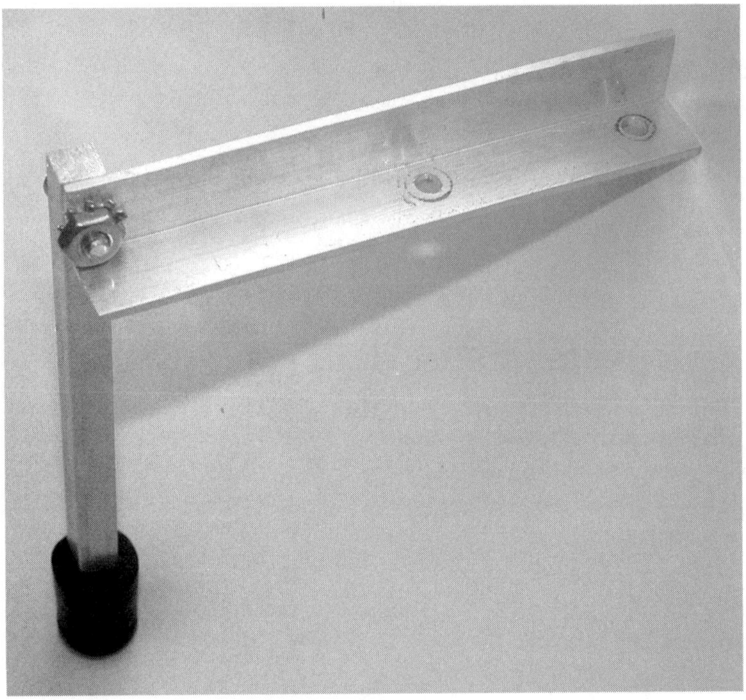

FIGURE 5.24 Back upper and lower leg assembly guide.

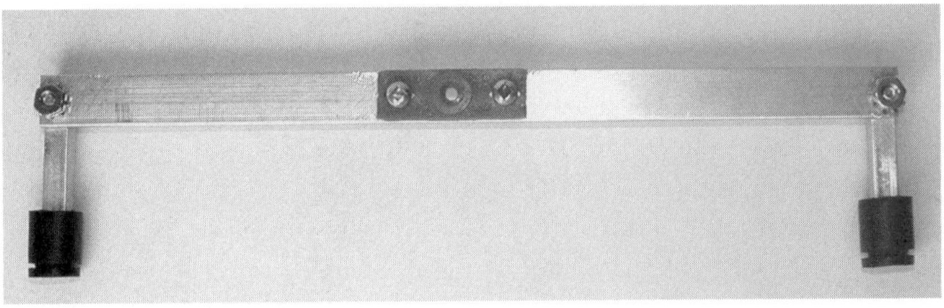

FIGURE 5.25 Middle leg assembled.

Centering the Servos

The next thing that needs to be done before attaching the legs to the servos is to center the servos. Take one of the servo horns supplied with the servo package and attach it to the servo so that it is in the middle point. Don't attach it with a screw. Rotate the servo by hand all the way counterclockwise and note where the horn is positioned.

Rotate the servo by hand all the way clockwise and note where it is positioned. If the position of the servo is not the same for the clockwise and counterclockwise position, then pull the servo horn off and rotate it by a few degrees and reattach to the servo. Repeat the above steps until the position of the horn is in the same mirrored position as the other side when fully rotated clockwise and counterclockwise, i.e., completely horizontal or vertical. Now adjust the servo by hand to the middle point and take the servo horn off. Repeat this procedure for all three servos. The servos' midpoint will be further calibrated electronically later on in the book.

Attaching the Legs to the Servos

Starting with the middle leg, line up the servo mount with the middle servo and secure with the servo mounting screw that came with the servo package. Use Figures 5.26 and 5.27 as a guide. Take the front left leg assembly made up of pieces S and M and secure it to the front left servo with the servo mounting screw. Take the front right leg assembly made up of pieces R and L and secure it to the front right servo with the servo mounting screw. Use Figures 5.28 and 5.29 as a guide.

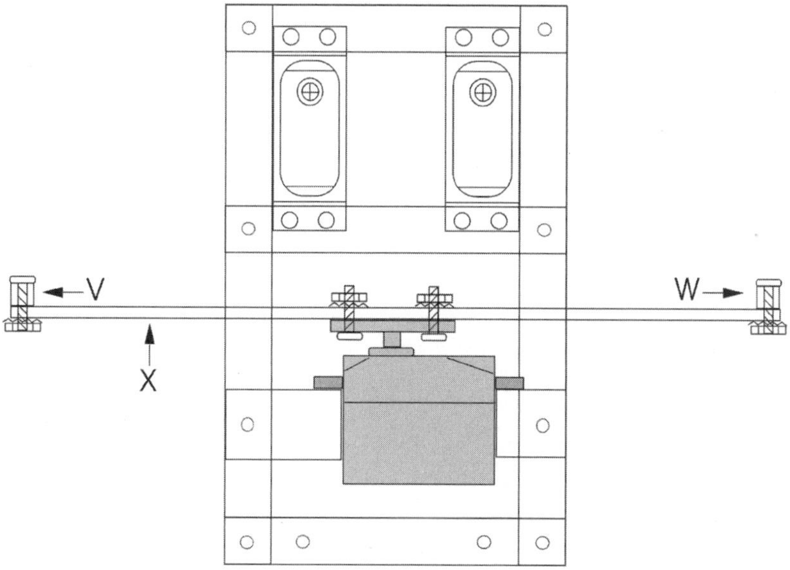

FIGURE 5.26 Middle leg attached to middle servo—top view.

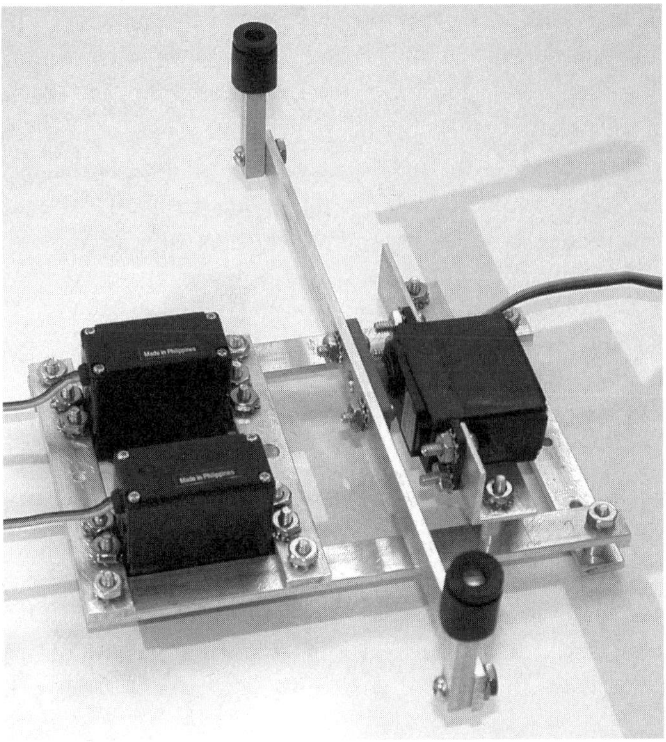

FIGURE 5.27 Middle leg attached to middle servo—underside view.

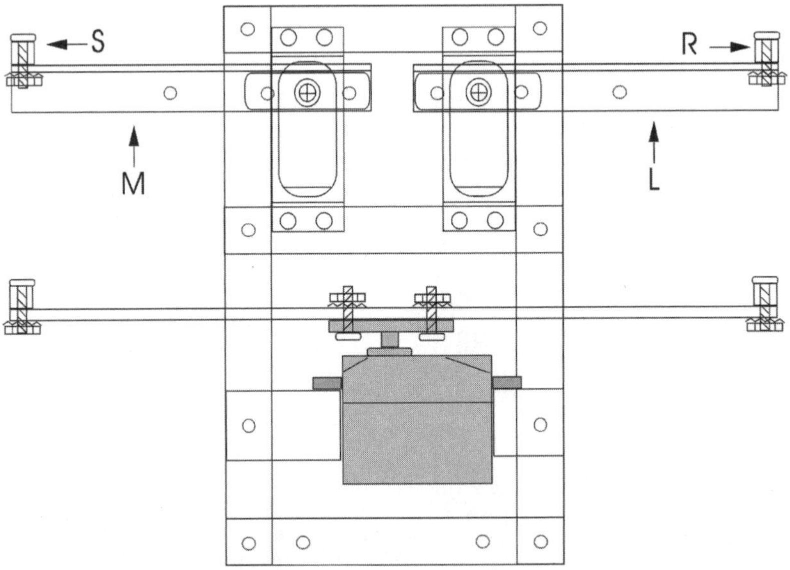

FIGURE 5.28 Front legs attached to front servos—top view.

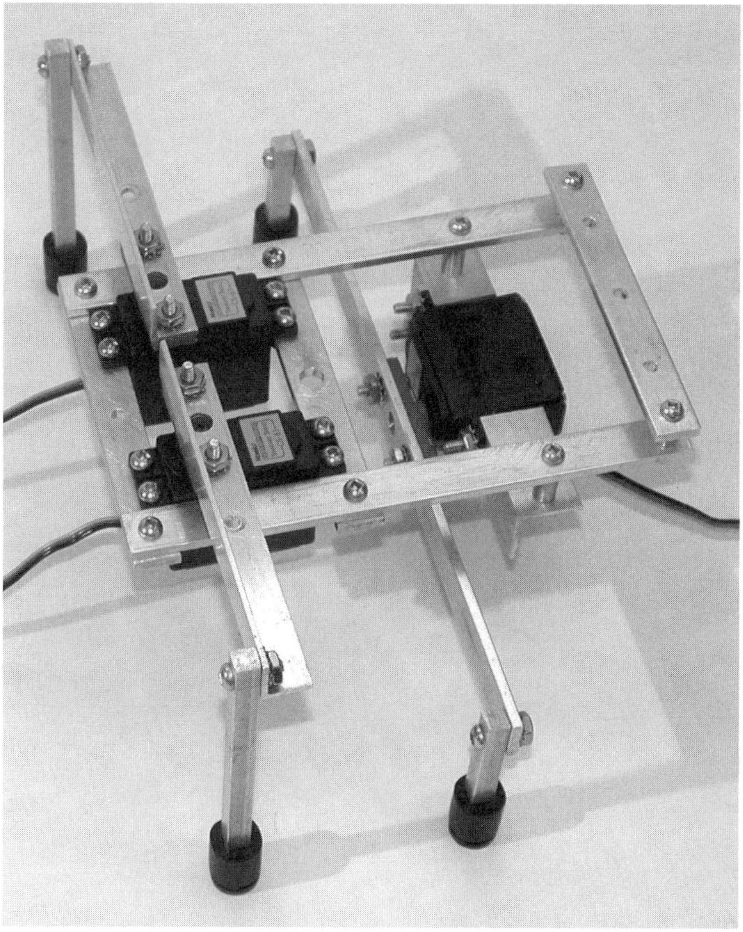

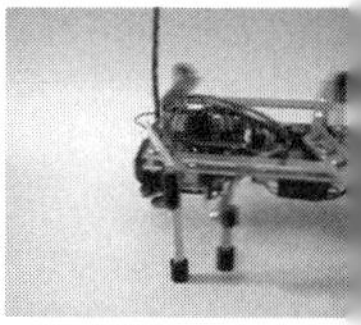

FIGURE 5.29 **Front legs attached to front servos—actual top view.**

Attaching the Back Legs to the Frame

Secure the back right leg assembly made up of pieces O and U to the rear right mount hole on piece C using a 6/32 × 1-inch machine screw, three nylon washers, and a 6/32-inch locking nut, as shown in Figures 5.30 and 5.31. Tighten the screw and lock nut with just enough torque so that the leg swivels without any resistance. The nylon washers form a simple bearing. Apply a small amount of white lithium grease to the top and bottom surfaces of each nylon washer before assembling. Follow Figures 5.30 and 5.31 as a guide. Use the exact same procedure to mount the left leg assembly made up of pieces N and T to the left leg mount hole on piece C. Figures 5.32 and 5.33 show the back legs attached to the frame.

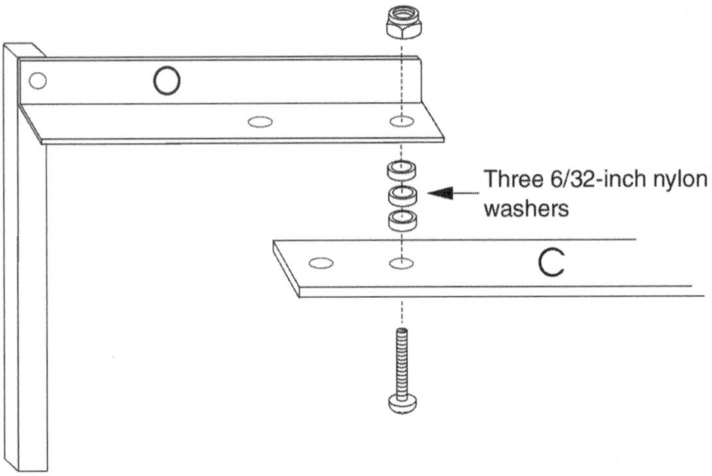

FIGURE 5.30 Back leg nylon washer bearing assembly.

FIGURE 5.31 Back leg bearing assembled on rear mounting hole.

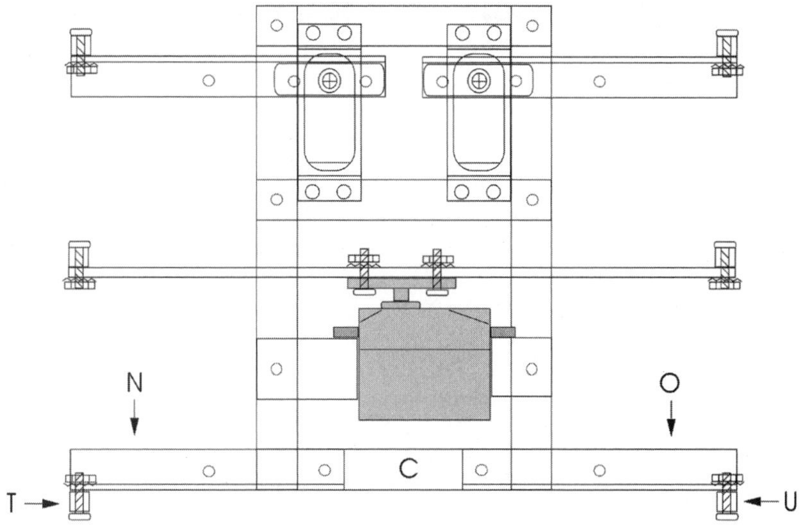

FIGURE 5.32 Back legs attached to rear frame mounting holes—top view.

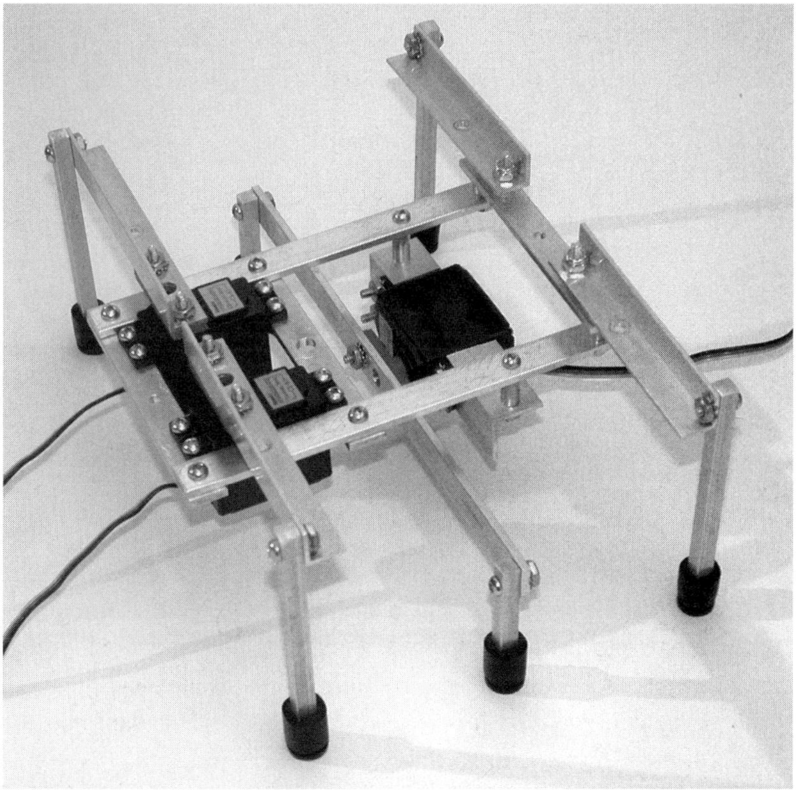

FIGURE 5.33 Back legs attached to frame.

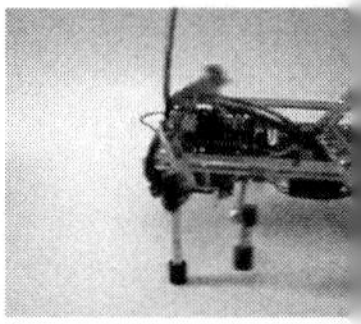

CONNECTING THE MECHANICAL LINKAGES

Fasten the mechanical linkage piece P to the under side of the left front and left back leg assemblies with a 6/32 × 1/2-inch machine screw and lock nut. Apply a small amount of white lithium grease around the holes of the linkage and the leg before assembling. Tighten the screw and lock nut with just enough torque so that the legs swivel together without any resistance. Use Figures 5.34 and 5.35 as a guide. Follow the same procedure for linkage Q with the right front and back legs.

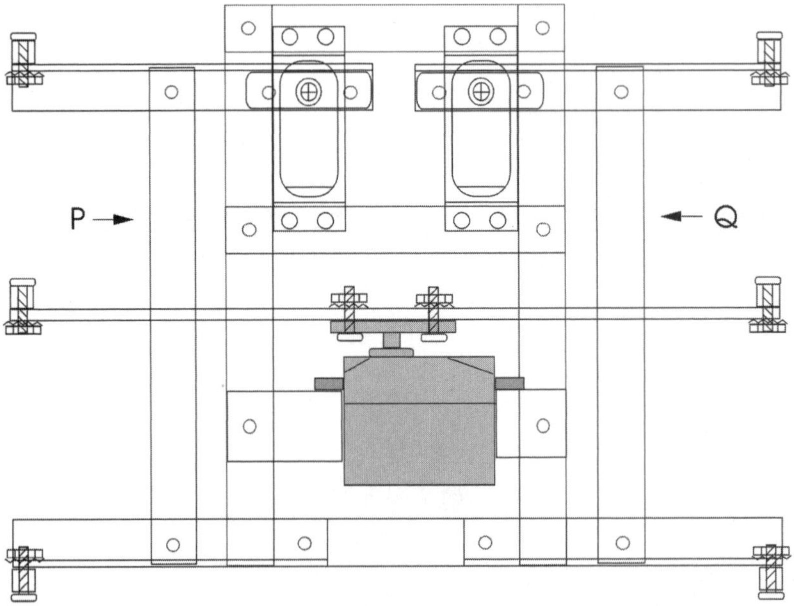

FIGURE 5.34 Mechanical linkages attached to front and back legs—top view.

Attaching the Feet and Screw Covers

Locate the six rubber feet and push one onto each leg so that there is a space of about 1/4 inch from the end of the leg to the end of the foot. Place the robot on a flat surface and adjust each foot up or down until the robot does not wobble. If you can't find the feet shown in the picture, then choose something similar from the hardware store. Cut four pieces of the 1/4-inch plastic wire protector to a length of 1 1/4 inches and two pieces 1/2 inch long. Place the 1 1/4-inch length wire protectors on the four outer legs at the top so that they cover the screws. Place the 1/2-inch pieces on the middle legs. Secure in place with hot glue. Figure 5.36 shows the rubber feet and screw covers in place.

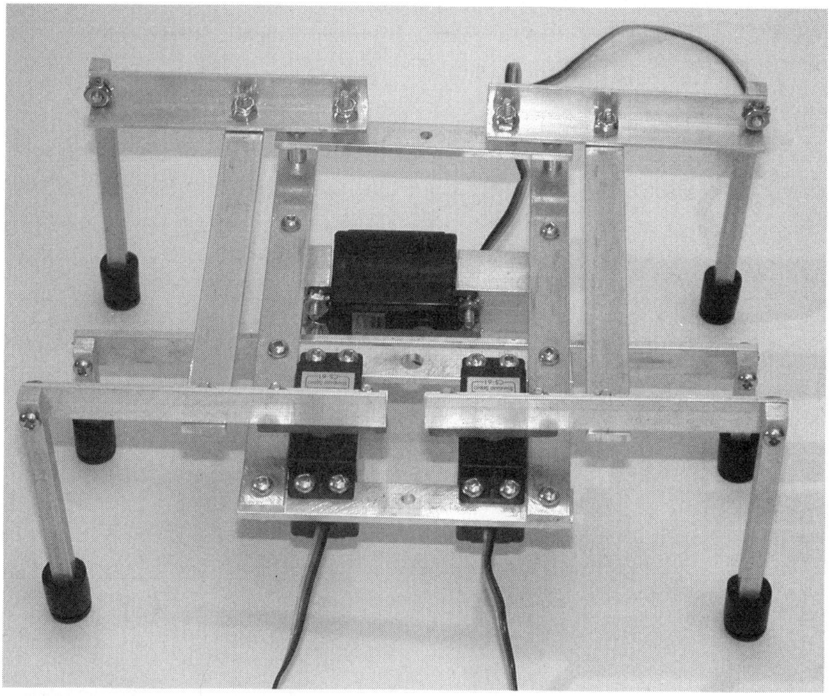

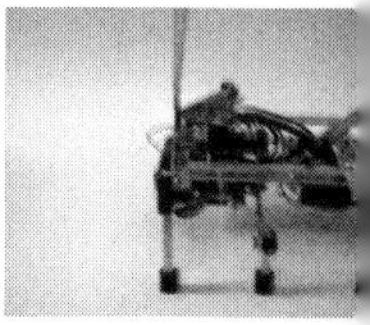

FIGURE 5.35 Mechanical linkages—front view.

FIGURE 5.36 Rubber feet and screw covers in place.

Summary

At this point, the basic mechanical structure of the robot is complete. In Chapter 6, we will construct the main controller board, which is designed around the PICmicro MCU 16F84 microcontroller. This computer on a chip will control all of the robot's functions.

6

THE MAIN CONTROLLER BOARD

This chapter deals with the construction of the robot's main controller circuit and the construction of the printed circuit board. Table 6.1 lists all of the parts necessary to build the controller board.

All the robot's functions are controlled by a Microchip™ PIC 16F84 microcontroller. The microcontroller is an entire computer on a chip. This makes it possible to eliminate a large amount of hardware that would otherwise be required. The microcontroller serves as the robot's "brain," controlling and managing all functions, sensors and reflexes. The 16F84 microcontroller we are using will be clocked at 4 MHz and operates on a 5-volt DC supply, produced from a 78L05 voltage regulator with the source being a 9V battery. The servos are powered by a separate 6-volt DC battery pack. The power supplies are separated to isolate noise from the servos and to keep a steady 5 volts to the processor when the 6V supply to the servos gets low. This enables the robot to operate for a much longer amount of time in between charges. As you can see from the schematic shown in Figure 6.1, the I/O lines are used as inputs and outputs. Some are used to receive signals from an infrared sensor board that acts as the robot's sense of sight. Others control the position of the robot's legs to achieve walking gaits, receive signals from a remote control, turn on two light-emitting diodes, and as an audio output to a piezo speaker. Each of the controller board's functions will be covered in detail in Chapters 9 and 10.

TABLE 6.1 Parts List for the Main Controller Board

PART	QUANTITY	DESCRIPTION
Semiconductors		
U1	1	78L05 5V regulator
U2	1	PIC 16F84 flash microcontroller mounted in socket
Q1	1	2N3904 NPN transistor
D1	1	Red light-emitting diode
D2	1	Green light-emitting diode
Resistors		
R1	1	4.7 KΩ 1/4-watt resistor
R2, R3, R4	3	1 KΩ 1/4-watt resistor
R5	1	100Ω 1/4-watt resistor
Capacitors		
C1	1	0.1 μf capacitor
C2, C3	2	22 ρf
Miscellaneous		
JP1 – JP5, JP8	6	3 post header connector–2.5 mm spacing
JP6, JP7	4	2 post header connector–2.5 mm spacing
S1	1	DPDT 3-position toggle switch
BT1,BT2	2	2-contact terminal block
Y1	1	4-MHz crystal
Piezo buzzer	1	Standard piezoelectric element
Battery holder	1	4-cell AA battery holder–6V output
Battery strap	2	9V-type battery strap
IC socket	1	18-pin IC socket–soldered to PC board U2

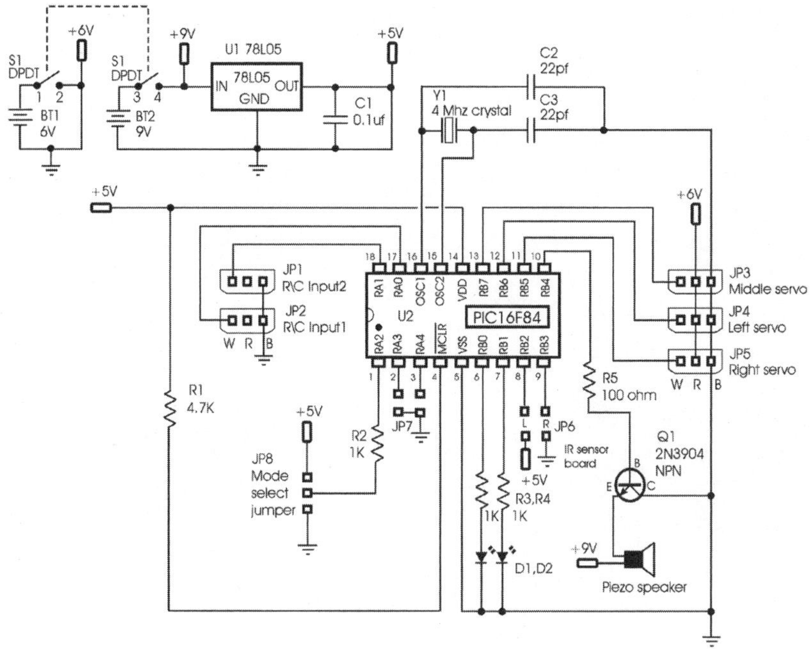

FIGURE 6.1 **Main controller board schematic.**

Printed Circuit Board Fabrication

The easiest way to produce quality printed circuit boards is by using the positive photo fabrication process. To fabricate the printed circuit board, photocopy the artwork in Figure 6.2 onto a transparency. The component placement is shown in Figure 6.3. Make sure that the photocopy is the exact size of the original. For convenience, you can choose to download the file from my Web site: http://www.thinkbotics.com and just print the file onto a transparency using a laser or inkjet printer with a minimum resolution of 600 dpi. You can also choose to buy the finished printed circuit board from the same Web site. After the artwork has been successfully transferred to a transparency, the following instructions can be used to create a board. A 4 × 6-inch pre-sensitized positive copper board is ideal. When you place the transparency on the copper board, it should be oriented so that it is exactly the same as it is in Figure 6.2. It would be a good idea to make the infrared sensor circuit board in Chapter 7 at the same time.

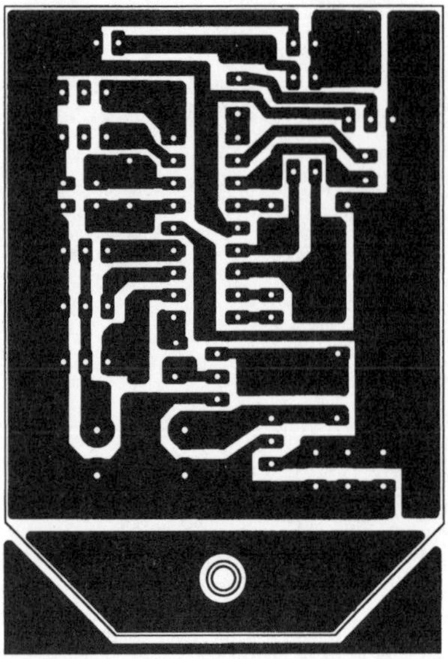

FIGURE 6.2 PCB foil pattern artwork.

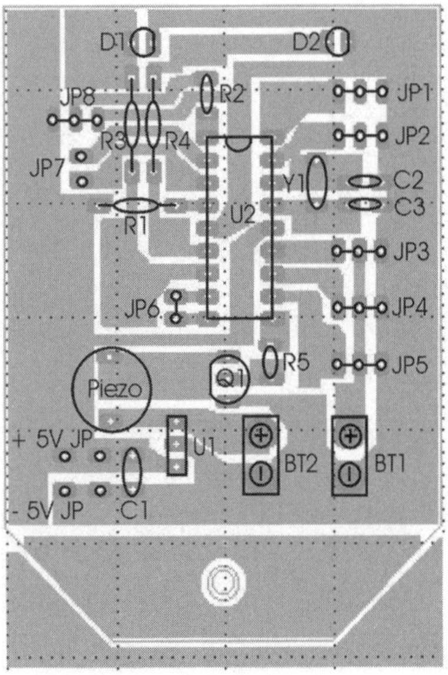

FIGURE 6.3 PCB component side parts placement.

A company that specializes in providing presensitized copper boards and all the chemistry needed to fabricate boards is M.G. Chemicals. Information on how to obtain all of the supplies can be found on their Web site: http://www.mgchemicals. com.

Figure 6.4 shows the developer, ferric chloride, and presensitized copper board that will be used for fabricating the circuit board.

FIGURE 6.4 Photo fabrication kit.

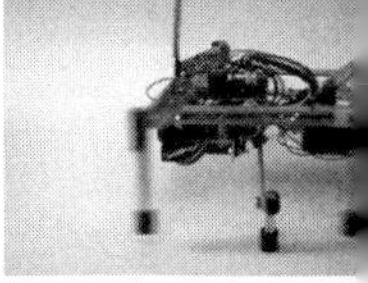

Follow the next five steps to make your own printed circuit board:

1. **Setup**—Protect surrounding areas from developer and other splashes that may cause etching damage. Plastic is ideal for this. Work under safe light conditions. A 40W incandescent bulb works well. Do not work under fluorescent light. Just prior to exposure, remove the white protective film from the presensitized board. Peel it back carefully.
2. **Exposing your board**—For best results, use the M.G. Chemicals cat. # 416-X exposure kit. However, any inexpensive lamp fixture that will hold two or more 18-inch fluorescent tubes is suitable.

   Directions: Place the presensitized board, copper side, toward the exposure source. Positive film artwork should be laid onto the "presensitized" copper side of the board and positioned as desired.

   Artwork should have been produced by a 600-dpi or better printer. A glass weight should then be used to cover the artwork, ensuring that no light will pass under the traces (approximately 3 mm glass thickness or greater works best). Use a 10-minute exposure time at a distance of 5 inches.

3. **Developing your board**—The development process removes any photoresist that was exposed through the film positive to ultraviolet light. **Warning: The developer contains sodium hydroxide and is highly corrosive. Wear rubber gloves and eye protection while using it. Avoid contact with eyes and skin. Flush thoroughly with water for 15 minutes if it is splashed in eyes or on the skin.**

   Directions: Using rubber gloves and eye protection, dilute one part M.G. Cat. # 418 developer with ten parts tepid water (weaker is better than stronger). In a plastic tray, immerse the board, copper side up, into the developer, and you will quickly see an image appear while you are lightly brushing the resist with a foam brush. This should be completed within one to two minutes. Immediately neutralize the development action by rinsing the board with water. The exposed resist must be removed from the board as soon as possible. When you are done with the developing stage, the only resist remaining will be covering what you want your circuit to be. The rest should be completely removed.

4. **Etching your board**—For best results, use the 416-E Professional Etching Process Kit or 416-ES Economy Etching Kit.

   The most popular etching matter is ferric chloride, M.G. Cat. # 415, an aqueous solution that dissolves most metals. Warning: This solution is normally heated up during use, generating unpleasant and caustic vapors; it is very important to have adequate ventilation. Use only glass or plastic containers. Keep out of reach of children. May cause burns or stain. Avoid contact with skin, eyes or clothing. Store in plastic container. Wear eye protection and rubber gloves.

   If you use ferric chloride cold, it will take a long time to etch the board. To speed up the etching process, heat up the solution. A simple way of doing this is to immerse the ferric chloride bottle or jug in hot water, adding or changing the water to keep it heating. A thermostat-controlled crock pot is also an effective way to heat ferric chloride as are thermostatically controlled submersible heaters—(glass enclosed, such as an aquarium heater). An ideal etching temperature is 50°C (120°F). Be careful not to overheat the ferric chloride. The absolute maximum working temperature is about 57°C (135°F). The warmer your etch solution, the faster your boards will etch. Ferric chloride solution can be used over and over again, until it becomes saturated with copper. As the solution becomes more saturated, the etching time will increase. Agitation assists in removing unwanted copper faster. This can be accomplished by using air bubbles from two aquarium air wands with an aquarium air pump. Do not use an aquarium "air stone." The etching process can be assisted by brushing the unwanted resist with a foam brush while the board is submerged in the ferric chloride. After the etching process is completed, wash the board thoroughly under running water. Do not remove the remaining resist protecting your circuit or image, as it protects the copper from oxidation. If you require it to be removed, use a solvent cleaner.

5. **Soldering your board**—Removal of resist is not necessary when soldering components to your board. When you leave the resist on, your circuit is protected from oxidation. Tin-plating your board is not necessary. In the soldering process, the heat disintegrates the resist underneath the solder, producing an excellent bond.

Drilling and Parts Placement

Use a 1/32-inch drill bit to drill all the component holes on the printed circuit board. Drill the holes for the terminal blocks (BT1 and BT2) and the voltage regulator (U1) with a 3/64-inch drill bit. Use Figures 6.1 and 6.3 to place the parts on the component side of the circuit board. Note that the PIC 16F84 microcontroller (U2) is mounted in an 18-pin IC socket. The 18-pin socket is soldered to the PC board, and the PIC is inserted after it has been programmed. Use a fine-toothed saw to cut the board along the guide lines and drill the mounting hole using a 6/32-inch drill bit. Figures 6.5 and 6.6 show the finished main controller board.

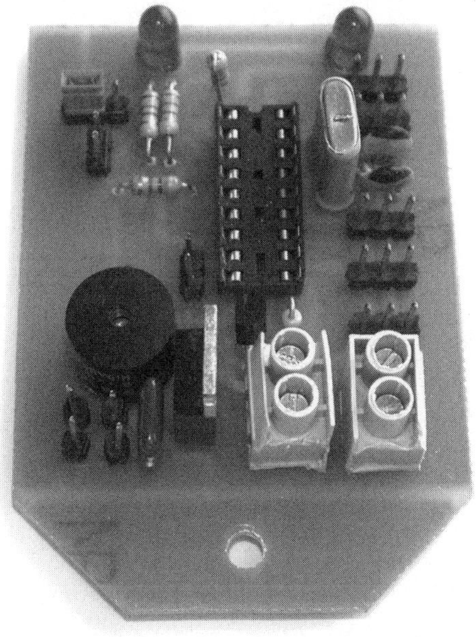

FIGURE 6.5 Parts soldered to the printed circuit board.

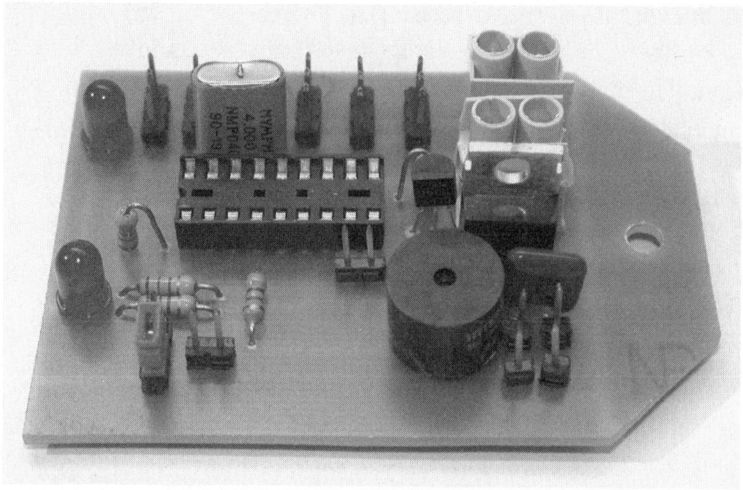

FIGURE 6.6 Finished main controller board.

Check the finished board for any missed or cold soldered connections and verify that all the components have been included. The board will be tested in Chapter 9 when programming the PIC microcontroller is covered.

Summary

Now that the main controller board is finished, it can be set aside until it is integrated with the chassis and other components in Chapter 8. Chapter 7 will focus on constructing the infrared sensor board that will be used by the robot to avoid obstacles.

THE INFRARED SENSOR BOARD

The infrared sensor board schematic is shown in Figure 7.1. It is comprised of two pairs of infrared light-emitting diodes and Panasonic PNA4602M IR sensor modules. The emitters and sensors are arranged on right angles, providing the robot with obstacle-avoidance capabilities to the front and on both sides of its head. This gives the robot a sense of vision that will also be used to map its environment for more advanced applications. Table 7.1 is a list of all the parts needed to construct the board.

The 555 timer in the circuit is used to modulate the infrared LEDs at a frequency determined by C1 and R3. R3 is an adjustable 10k potentiometer that will be used to find the optimum frequency during calibration. In our application, we will use a frequency between 38 and 42 kHz, so that a more meaningful signal will be sent from the PNA4602 sensor module to the microprocessor.

The PNA4602M shown in Figure 7.2 is designed to detect only infrared radiation that is modulated at 38 kHz and rejects all other light sources. This makes the module an ideal sensor for daylight conditions. The features include an extension distance of 8 meters or more. No external parts are required, and a resin filter makes the module unsusceptible to visible light. Table 7.2 lists the PNA4602M module's main characteristics. The output signals from the module will be processed and filtered by the microprocessor with a software routine discussed in Chapter 10.

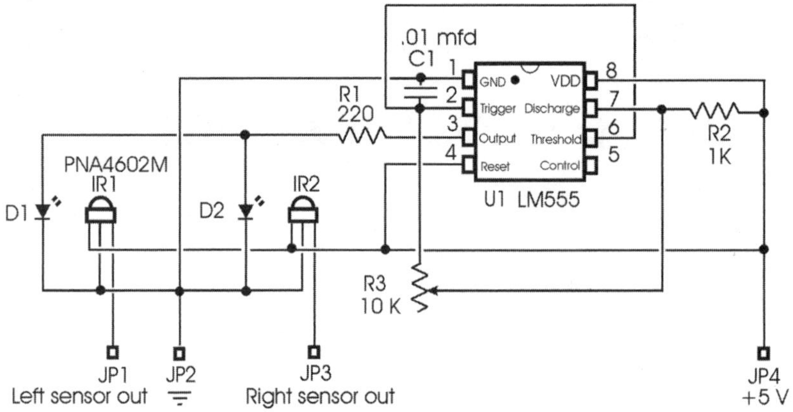

FIGURE 7.1 Infrared head sensor schematic.

TABLE 7.1 Parts List for Infrared Head Sensor

Part	Quantity	Description
Semiconductors		
U1	1	LM 555 timer integrated circuit
IR1, IR2	2	Panasonic PNA4602M infrared detector modules
D1, D2	2	Infrared light-emitting diodes
Resistors		
R1	1	220Ω 1/4-watt resistor
R2	1	1 KΩ 1/4-watt resistor
R3	1	10 KΩ adjustable potentiometer
Capacitors		
C1	1	.01 μfd capacitor
Miscellaneous		
JP1-JP4	2	2 post header connector
Printed circuit board	1	See details in chapter
IC socket	1	8-pin I.C. socket soldered to PC board to mount U1

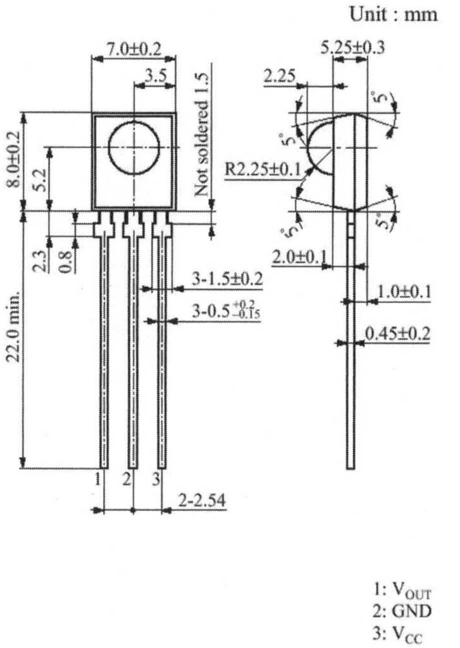

FIGURE 7.2 Diagram of PNA4602M infrared sensor module.

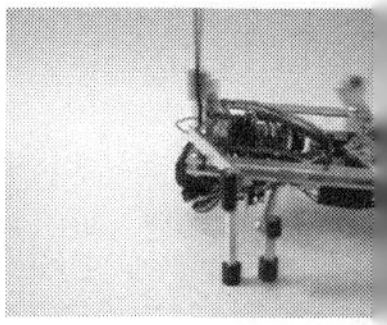

TABLE 7.2 Main Characteristics of the PNA4602M Module

PARAMETER	SYMBOL	MINIMUM	TYPICAL	MAXIMUM	UNIT
Operating supply voltage	Vcc	4.7	5.0	5.3	V
Current consumption	Icc	1.8	2.4	3.0	mA
Max. reception distance	Lmax	8	10		m
Low-level output voltage	Vol		0.35	0.5	V
High-level output voltage	Voh	4.8	5.0	Vcc	V
Low-level pulse width	Twl	200	400	600	μs
High-level pulse width	Twh	200	400	600	μs
Carrier frequency	Fo		38.0		kHz

Figure 7.3 illustrates the concept of infrared radiation, produced by the light-emitting diode, and being reflected from the surface of a solid object back to the infrared sensor module. Depending on the proximity of the object to the sensor, a greater number of infrared pulses will be reflected back. The number of reflected "hits" that the sensor receives in a given time frame allows the robot to determine how close it is to objects.

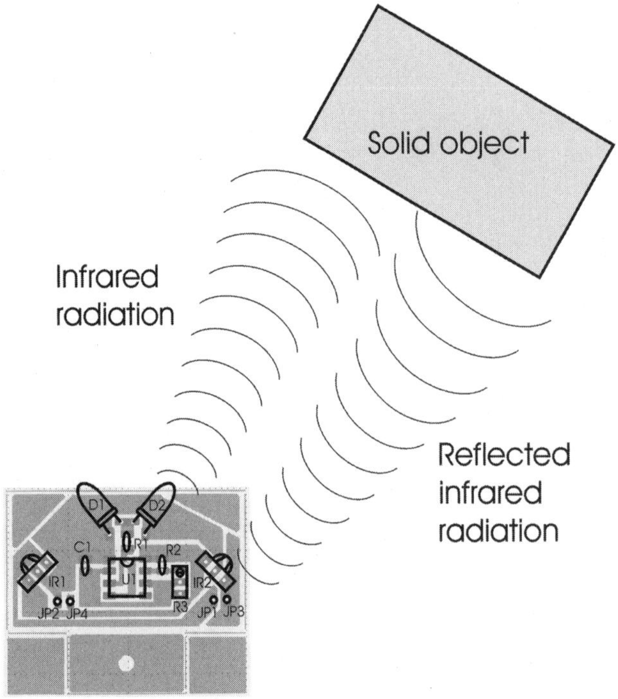

FIGURE 7.3 **Diagram of reflected infrared radiation.**

Constructing the Printed Circuit Board

To fabricate the printed circuit board, photocopy the artwork in Figure 7.4 onto a transparency. Make sure that the photocopy is the exact size of the original. For convenience, you can choose to download the file from my Web site, (http://www.thinkbotics.com/) and just print the file onto a transparency using a laser or inkjet printer with a minimum resolution of 600 dpi. You can also choose to buy the finished printed circuit board from my Web site. After the artwork has been successfully transferred to the transparency, follow the instructions in the previous chapter to create a circuit board. A 4 × 6-inch presensitized positive copper board is ideal. When the transparency is placed on the copper board, make sure that it is oriented exactly as shown in Figure 7.4.

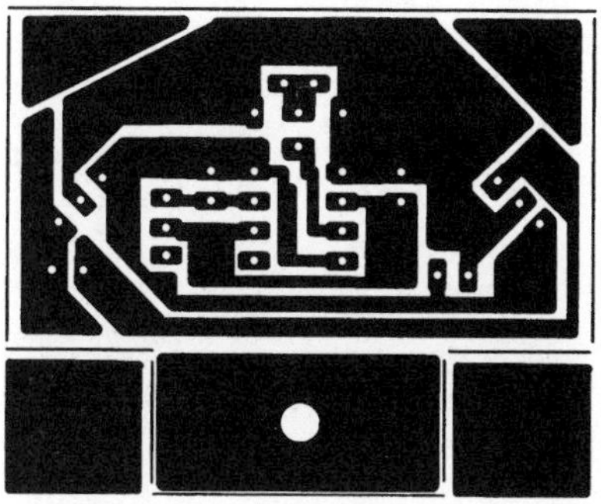

FIGURE 7.4 Infrared sensor module PCB foil pattern artwork.

DRILLING AND PARTS PLACEMENT

Use a 1/32-inch drill bit to drill all the component holes on the printed circuit board. Use Figures 7.5 and 7.1 to place the parts on the component side of the circuit board. Note that the 555 timer I.C. (U1) is mounted in an 8-pin I.C. socket. The 8-pin socket is soldered to the PC board and the LM555 is inserted into the socket. Use a fine-toothed saw to cut the board along the guide lines and drill the mounting hole using a 6/32-inch drill bit. Figure 7.6 shows the finished infrared sensor board.

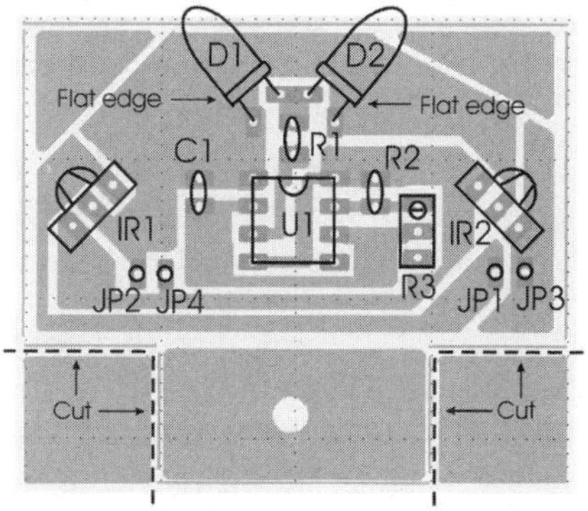

FIGURE 7.5 PCB component side parts placement.

FIGURE 7.6 Finished infrared sensor board.

Calibrating the Infrared Sensor Board

To calibrate the infrared sensors, a dual channel oscilloscope such as the one pictured in Figure 7.7 will be used. This method will be discussed in detail in this chapter. An alternate method that will also be covered is the use of a digital multimeter with frequency counting capabilities. The nice thing about having access to an oscilloscope is the visual aspect of measurement. Because our circuit is being analyzed by the oscilloscope, you will be able to see the reflected pulses of infrared radiation received by the infrared sensor modules.

Attach a 5-volt DC power source to jumper 4 with the ground attached to jumper 2. Connect the oscilloscope probe for channel 1 to R1 and channel 2 to jumper 3 using Figure 7.8 as a guide. The ground terminal of the oscilloscope should be attached to the ground connector of the circuit (jumper 2). Set the oscilloscope to a 2-volt per division scale and a time frame of 100 μ seconds.

Once the probes have been attached and the power turned on, you should see a square wave on channel 1. Adjust R3 until a frequency somewhere between 22.5 and 23.0 kHz is displayed on the oscilloscope. This frequency will only be used for testing the circuit at this time. If your oscilloscope does not have an automatic frequency measuring capability, use the formula: frequency = 1/time, where time is the period of the waveform. Place the circuit board on a table with nothing in front of the right infrared light-emitting diode and sensor module. Channel 2 should show no activity or a few pulses, as shown in Figure 7.9.

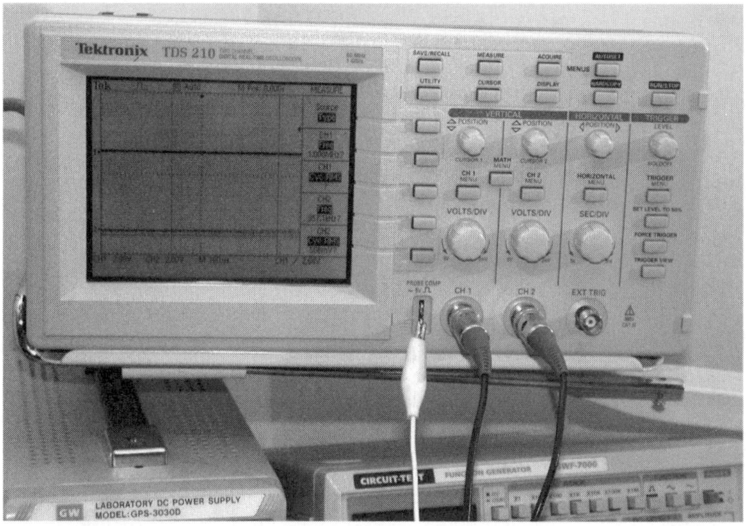

FIGURE 7.7 Dual channel digital oscilloscope.

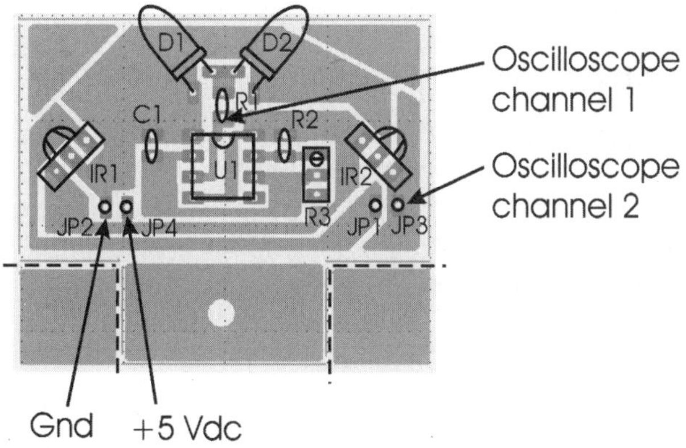

FIGURE 7.8 Oscilloscope probe placement.

Place a solid object directly in front and to the right of the module at a distance of 2 inches. A waveform resembling that shown in Figure 7.10 should be displayed showing all of the infrared pulses being reflected back. If the trace shows no change in activity, slowly rotate potentiometer R3 in very small increments to the left until activity takes place. If there is still nothing displayed on channel 2, slowly rotate potentiometer R3 to the right in very small increments until a waveform similar to that in Figure 7.10 is displayed.

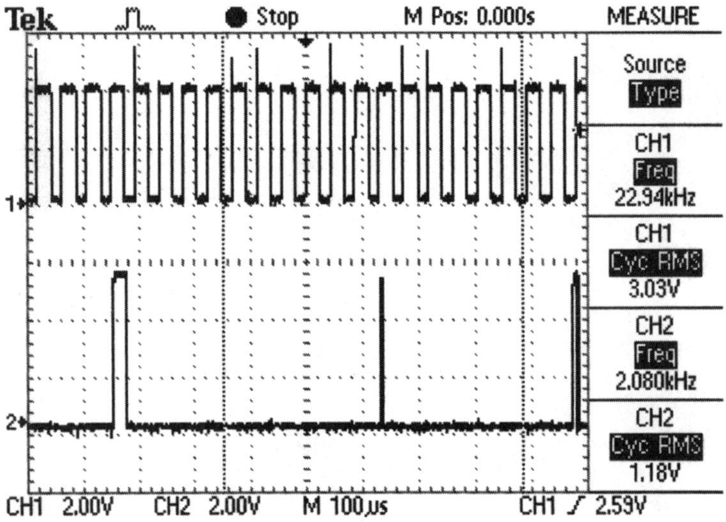

FIGURE 7.9 Waveform displayed when no object is present.

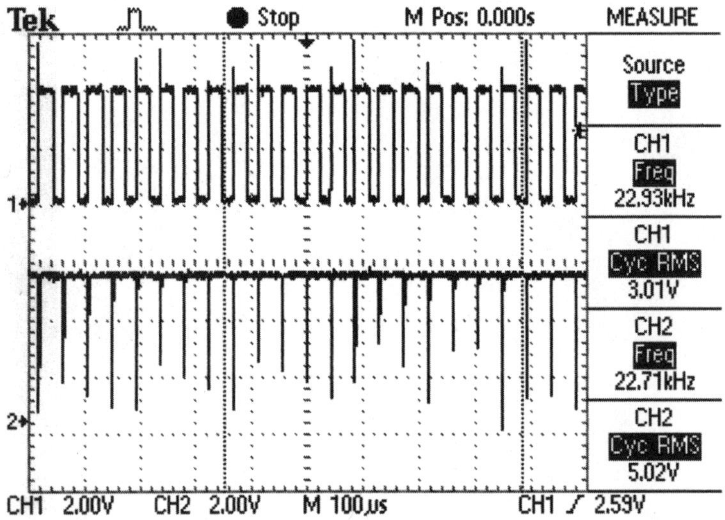

FIGURE 7.10 Waveform displayed when object is very close.

Move the solid object approximately 6 inches from the module. As the object moves further away, the number of reflected pulses seen on channel 2 of the oscilloscope will be smaller. The waveform should look similar to that shown in Figure 7.11.

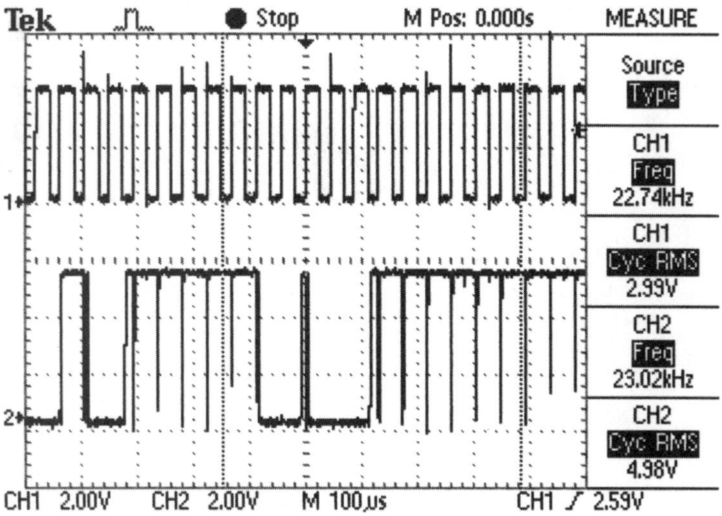

FIGURE 7.11 Waveform displayed when object is near.

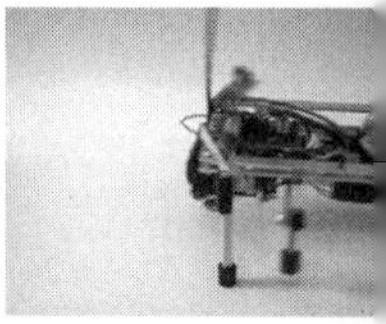

Now switch oscilloscope probe 2 to JP1 and repeat the above procedure on the left IR module without adjusting the frequency.

If all goes well with the previous procedure, adjust potentiometer R3 until a stable frequency between 40.9 and 41.6 kHz is measured. The optimum frequency seems to be 40.9 kHz.

CALIBRATING THE CIRCUIT USING A DIGITAL MULTIMETER

A digital multimeter with frequency counting capabilities like the one shown in Figure 7.12 will be needed to calibrate the infrared sensor board for the next method that will be described.

Figure 7.13 shows where the positive probe of the multimeter is connected. The common lead of the multimeter is connected to the ground or negative pin.

Turn on the multimeter and set it to measure frequency. Connect a 5-volt DC source to the circuit as shown in Figure 7.13. Adjust potentiometer R3 until a frequency of approximately 40.9 kHz is measured. For now, that is all that needs to be done using this method. The sensor board will be completely tuned using a software routine in Chapter 10.

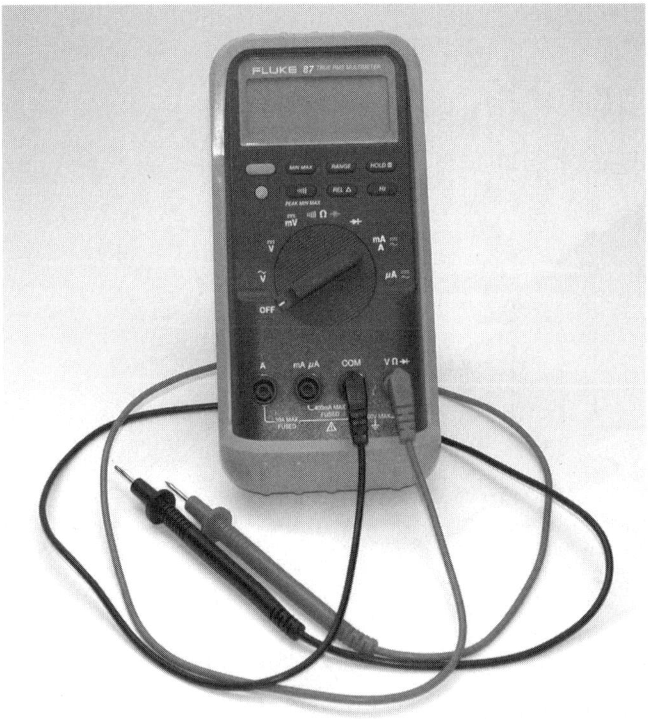

FIGURE 7.12 Digital multimeter with frequency measuring capabilities.

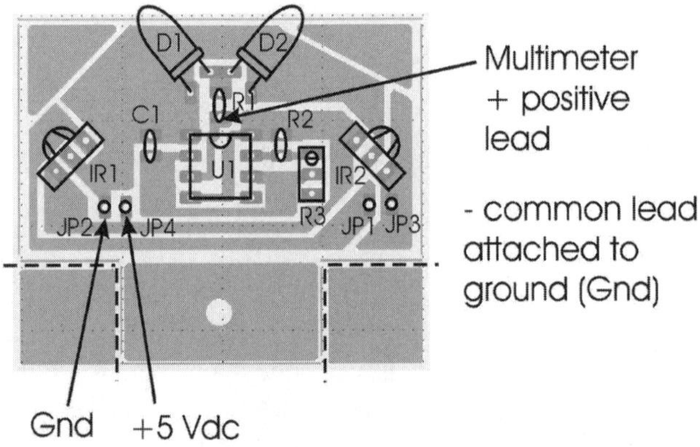

FIGURE 7.13 Multimeter probe connection guide.

Summary

With the completion of the infrared sensor board, the robot will possess obstacle-avoidance capabilities necessary for survival. The completed sensor board will be added to the robot chassis along with the main controller and power supply in Chapter 8.

8

FINAL ASSEMBLY

Now that the Insectronic's body, main controller, and infrared sensor boards are finished, it's time to put them all together. Table 8.1 is the list of parts needed for the final assembly. The first step is to mount the 6-volt battery pack and the 9-volt battery to the robot. This is accomplished by fabricating the battery holders as detailed in Figure 8.1 using 1/16 inch-thick aluminum. Cut the pieces and then drill with a 5/32-inch drill bit. Measure and mark the locations where the aluminum needs to be bent on 90-degree angles. Use a workbench vise or the edge of a table to bend the pieces.

The finished battery holders are illustrated in Figure 8.2. The battery holders are mounted to the robot's body along with the circuit boards as shown in Figures 8.3 and 8.4. Use 6/32 × 1/2-inch machine screws, lock washers, and nuts to fasten the battery holders and circuit boards in place. The 9-volt battery holder and head sensor are attached to part A (see Figures 5.3 and 5.4 in Chapter 5). The 6-volt battery pack holder and controller board are attached to part C (see Figures 5.3 and 5.4 in Chapter 5).

TABLE 8.1 Parts List for Final Assembly

PART	QUANTITY	DESCRIPTION
S1	1	DPDT 3-position toggle switch
Battery holders	2	See fabrication details in this chapter
Battery pack	1	4-cell AA battery holder—6V output
Battery strap	2	9V-type battery strap
Wire	2 feet	18 gauge connecting wire
9-volt battery	1	Power for main controller and I.R. sensor board
AA batteries	4	For 6V battery pack to power the servos
Header connector	4	2-post female header connector—2.5-mm spacing

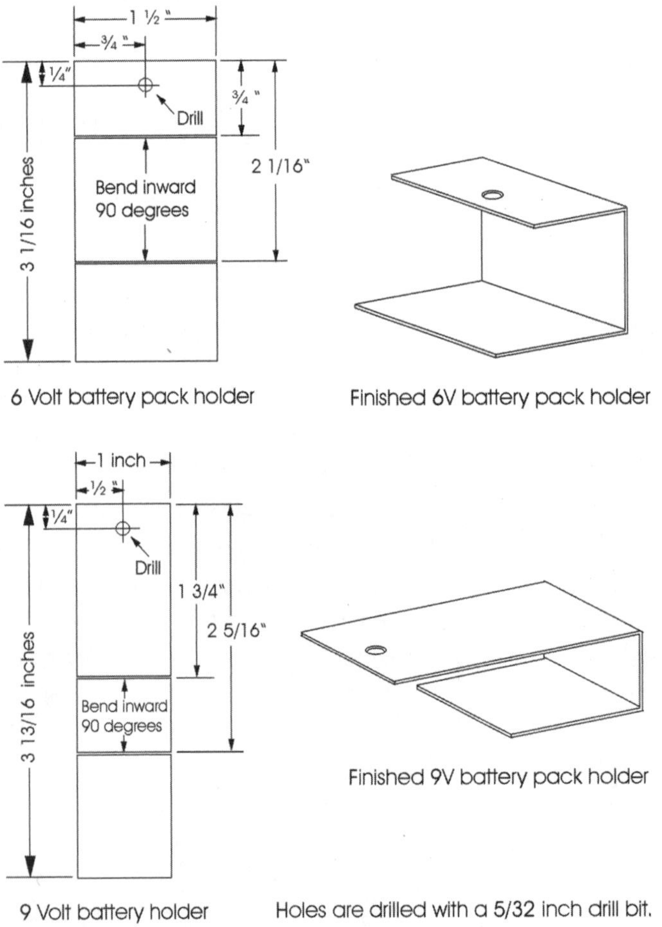

FIGURE 8.1 Cutting, drilling, and bending guide for battery holders.

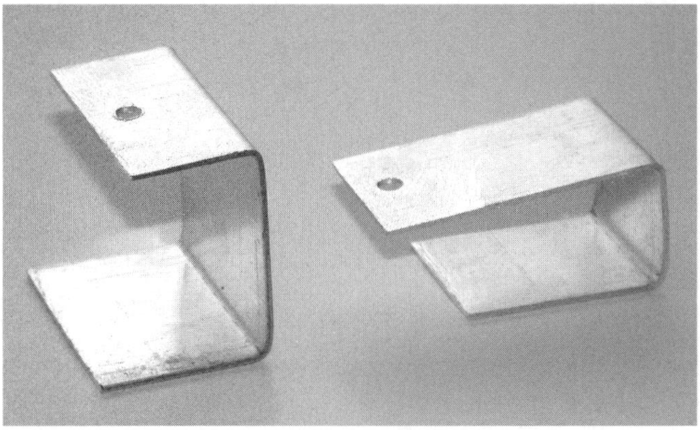

FIGURE 8.2 Finished battery holders.

FIGURE 8.3 I.R. head sensor and 9-volt battery holder—underside view.

FIGURE 8.4 Controller board and 6-volt battery pack holder attached to robot.

The next step is to mount the *double pole double throw* (DPDT) switch to the Insectronic's body. Figure 8.5 shows the switch mounted in the 1/4-inch hole drilled in piece B (see Figure 5.2 in Chapter 5).

FIGURE 8.5 DPDT switch fastened to part B.

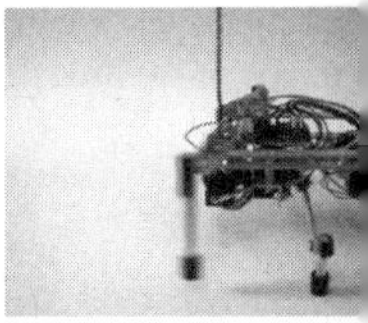

Follow the wiring diagram in Figure 8.6 to connect the various components. Start by wiring the positive (red) leads of the battery clips to the switch. Cut two 5-inch lengths of connection wire and strip the ends. Solder these to the switch as indicated and connect the other ends of the wires to the terminal block of the controller board as shown in the diagram. The negative (black) lead from the 6V battery pack can be screwed into the terminal block of the controller board as shown in the diagram. The negative lead from the 9V battery will need to be extended by 4 inches to reach the terminal block. Solder a 4-inch piece of wire to the lead and use heat shrink tubing or electrical tape to cover the connection and then attach to the terminal block. Cut four 9-inch lengths of connector wire and fabricate two jumper connectors consisting of two wires and two female 2-post header connectors each as shown in Figure 8.7. Use these to connect the infrared sensor board to the main controller as indicated in Figure 8.6.

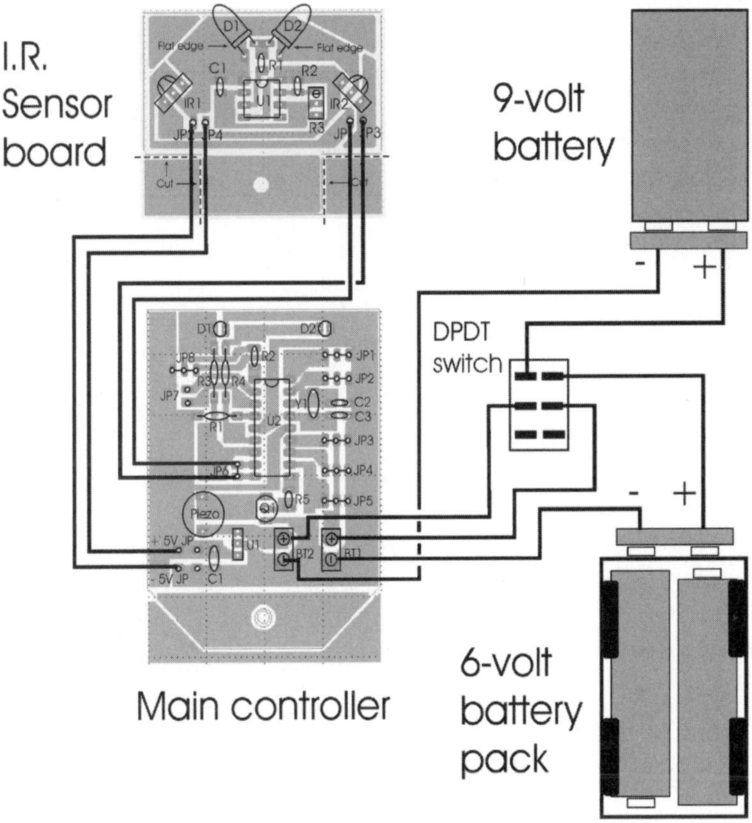

FIGURE 8.6 Final assembly wiring diagram.

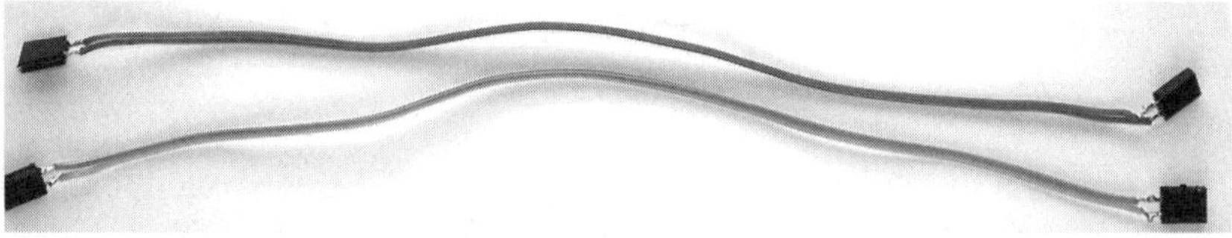

FIGURE 8.7 Connector wires.

Connect the servos to the main board by following Figure 8.8. The diagram refers to the left and right servo from the top perspective with the servos being located at the front of the robot. Place a 9V battery in the battery holder at the front of the robot underneath the infrared sensor board and connect the 9V battery strap. Insert four AA batteries in the 6V battery pack. Clip the battery pack into the rear battery pack holder and connect the 6V battery strap.

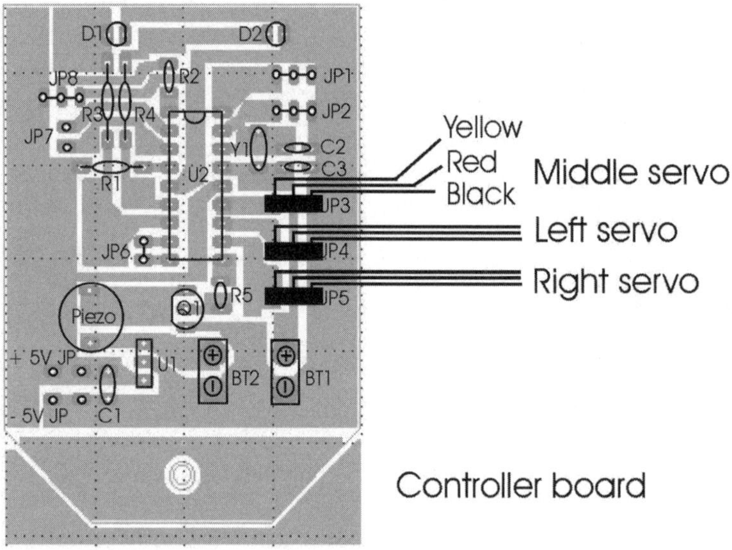

FIGURE 8.8 Servo connection diagram.

Now that the electronics and mechanics are in place, the next step is to calibrate the servos and electronics and program the Insectronic's brain. The robot should now look like the one pictured in Figure 8.9.

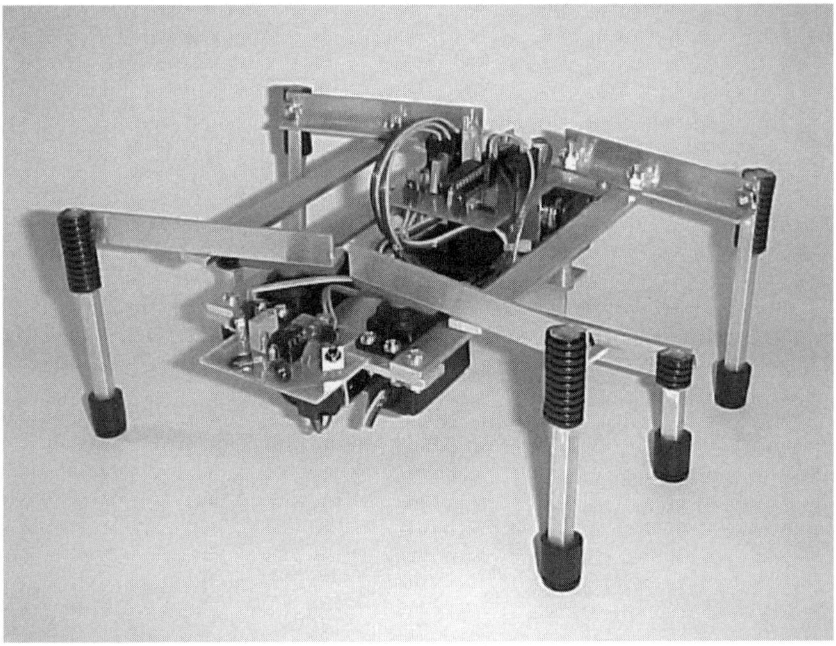

FIGURE 8.9 **Insectronic with circuit boards attached and wiring completed.**

Summary

At this stage, all of the individual components have been integrated. The robot is now a complete functioning system that can be programmed to walk and react to its environment. All that is needed is the addition of the 16F84 PICmicro. In Chapter 9, we will discuss microcontroller programming, and how this devise will bring life to our robot.

9

PIC PROGRAMMING

Microcontrollers

The microcontroller is an entire computer on a single chip. The advantage of designing around a microcontroller is that a large amount of electronics needed for certain applications can be eliminated. This makes it the ideal device for use with a small, lightweight, walking robot. The microcontroller is popular because the chip can be reprogrammed easily to perform different functions, and is very inexpensive. The microcontroller contains all the basic components that make up a computer. It contains a central processing unit, read-only memory, random-access memory, arithmetic logic unit, input and output lines, timers, serial and parallel ports, digital-to-analog converters, and analog-to-digital converters. The scope of this book is to discuss the specifics of how the microcontroller can be used as the processor for a walking robot. There are some good books on the market that go into detail about the internal functioning of the microcontroller. One such book, by John Iovine, is the *PIC Microcontroller Project Book*.

PIC 16F84 MCU

Microchip technology has developed a line of reduced instruction set computer (RISC) microprocessors called the programmable interface controller (PIC). The PIC uses what is known as "Harvard architecture." Harvard uses two memories and separate busses. The first memory is to store the program and the other is to store data. The advantage of this design is that instructions can be fetched by the CPU at the same time that RAM is being accessed. This greatly speeds up execution time. The architecture commonly used for most computers today is known as Von Neumann architecture. This design uses the same memory for control and RAM storage, and slows down processing time.

We will be using the PIC16F84, shown in Figure 9.1, as the processor for Insectronic. This device can be reprogrammed over and over because it uses flash read-only memory for program storage. This makes it ideal for experimenting because the chip does not need to be erased with an ultraviolet light source every time you need to tweak the code or try something new.

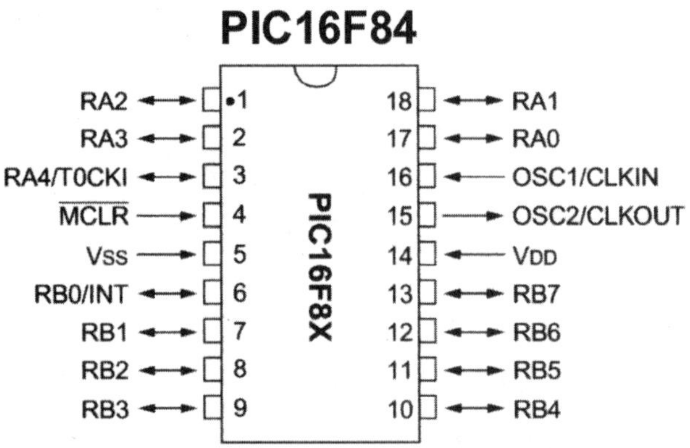

FIGURE 9.1 **Pinout of the PIC 16F84 microcontroller.**

The PIC 16F84 is an 18-pin device with an 8-bit data bus and registers. We will be using a 4-MHz crystal for the clock speed. This is very fast for our application when you consider that it is running machine code at 4 million cycles per second. The PIC 16F84 is equipped with two I/O ports, port A and port B. Each port has two registers associated with it. The first register is the TRIS (Tri State) register. The value loaded into this register determines if the individual pins of the port are treated as inputs or outputs. The other register is the address of the port itself. Once the ports have been configured using the TRIS register, data can then be written or read to the port using the port register address.

Port B has eight I/O lines available and Port A has five I/O lines. Insectronic will be using all eight I/O lines of Port B and three lines of Port A as shown in Figure 9.2.

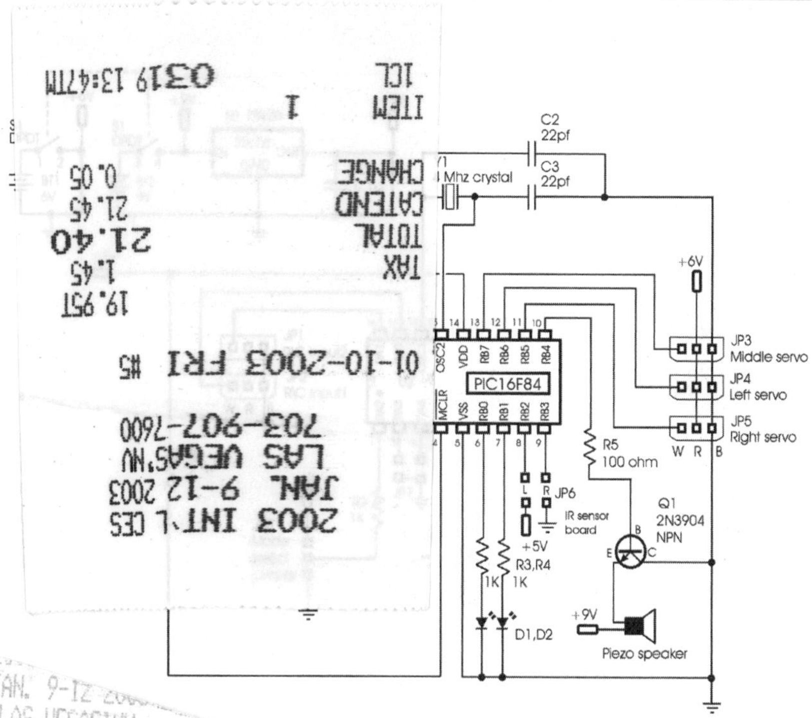

robot's main controller board schematic.

how the various pins of Port A and Port B will be used as inputs
trol the different functions of the robot.

16F84 Port A and B Connection Table

	ONFIGURATION	ROBOT CONNECTION
RB0	Output	Left light-emitting diode
RB1	Output	Right light-emitting diode
RB2	Input	Left input from infrared sensor board
RB3	Input	Right input from infrared sensor board
RB4	Output	Piezoelectric buzzer
RB5	Output	Right servo
RB6	Output	Left servo
RB7	Output	Middle servo

continued on next page

TABLE 9.1 PIC 16F84 Port A and B Connection Table (continued)

PORT A	CONFIGURATION	ROBOT CONNECTION
RA0	Input	Radio control input 1
RA1	Input	Radio control input 2
RA2	Input	Mode select jumper
RA3	Input/output	Calibration and experimentation
RA4	Input/output	Later experimentation

PICBASIC PRO COMPILER

Microengineering Labs develops the PicBasic Pro Compiler shown in Figure 9.3. It is a programming language that makes it quick and easy to program Microchip Technology's powerful PICmicro microcontrollers. It can be purchased from Microengineering Labs whose Web site is located at http://www.microengineeringlabs.com/.

FIGURE 9.3 PicBasic Pro Compiler.

The BASIC language is much easier to read and write than Microchip assembly language, and will be used to program Insectronic. The PicBasic Pro Compiler is "BASIC Stamp II like" and has most of the libraries and functions of both the BASIC Stamp I and II. Because it is a true compiler, programs execute much faster and may be longer than their Stamp equivalents.

One of the advantages of the PicBasic Pro compiler is that it uses a real IF..THEN..ELSE..ENDIF instead of the IF..THEN(GOTO) of the Stamps. These and other differences are spelled out in the PBP manual.

PBP defaults to create files that run on a PIC16F84-04/P clocked at 4 MHz. Only a minimum of other parts are necessary: two 22pf capacitors for the 4-MHz crystal, a 4.7K pull-up resistor tied to the /MCLR pin, and a suitable 5-volt power supply. Many PICmicros other than the 16F84, as well as oscillators of frequencies other than 4 MHz, may be used with the PicBasic Pro Compiler.

The PicBasic Pro Compiler produces code that may be programmed into a wide variety of PICmicro microcontrollers having from 8 to 84 pins and various on-chip features including A/D converters, hardware timers, and serial ports. For general-purpose PICmicro development using the PicBasic Pro Compiler, the PIC16F84, 16F876, and 16F877 are the current PICmicros of choice. These microcontrollers use flash technology to allow rapid erasing and reprogramming to speed program debugging. With the click of the mouse in the programming software, the flash PICmicro can be instantly erased and then reprogrammed again and again. Other PICmicros in the 12C67x, 14C000, 16C55x, 16C6xx, 16C7xx, 16C9xx, 17Cxxx, and 18Cxxx series are either *one-time programmable* (OTP) or have a quartz window in the top (JW) to allow erasure by exposure to ultraviolet light for several minutes. The PIC16F84 and 16F87x devices also contain between 64 and 256 bytes of non-volatile data memory that can be used to store program data and other parameters even when the power is turned off. This data area can be accessed simply by using the PicBasic Pro Compiler's READ and WRITE commands. (Program code is always permanently stored in the PICmicro's code space whether the power is on or off.)

By using a flash PICmicro for initial program testing, the debugging process may be sped along. Once the main routines of a program are operating satisfactorily, a PICmicro with more capabilities or expanded features of the compiler may be utilized.

Software Installation

The PicBasic Pro files are compressed into a self-extracting file on the diskette. They must be uncompressed to your hard drive before use. To uncompress the files, create a subdirectory on your hard drive called PBP or another name of your choosing by typing:

```
md PBP
```

at the DOS prompt. Change to the directory:

```
cd PBP
```

Assuming the distribution diskette is in drive a:, uncompress the files into the PBP subdirectory:

```
a:\pbpxxx -d
```

where `xxx` is the version number of the compiler on the disk. Don't forget the `-d` option on the end of the command. This ensures that the proper subdirectories within PBP are created.

Make sure that FILES and BUFFERS are set to at least 50 in your `CONFIG.SYS` file. Depending on how many FILES and BUFFERS are already in use by your system, allocating an even larger number may be necessary.

See the `README.TXT` file on the diskette for more information on uncompressing the files. Also read the `READ.ME` file that is uncompressed to the PBP subdirectory on your hard drive for the latest PicBasic Pro Compiler information. Table 9.2 is a list of the different PicBasic Pro compiler statements that are available to the PIC software developer.

TABLE 9.2 PicBasic Pro Statement Reference

STATEMENT	DESCRIPTION
@	Insert one line of assembly language code.
ADCIN	Read on-chip analog to digital converter.
ASM..ENDASM	Insert assembly language code section.
BRANCH	Computed GOTO (equiv. to ON..GOTO).
BRANCHL BRANCH	out of page (long BRANCH).
BUTTON	Debounce and autorepeat input on specified pin.
CALL	Call assembly language subroutine.
CLEAR	Zero all variables.
CLEARWDT	Clear (tickle) Watchdog Timer.

continued on next page

TABLE 9.2 PicBasic Pro Statement Reference (continued)

STATEMENT	DESCRIPTION
COUNT	Count number of pulses on a pin.
DATA	Define initial contents of on-chip EEPROM.
DEBUG	Asynchronous serial output to fixed pin and baud.
DEBUGIN	Asynchronous serial input from fixed pin and baud.
DISABLE	Disable ON DEBUG and ON INTERRUPT processing.
DISABLE DEBUG	Disable ON DEBUG processing.
DISABLE INTERRUPT	Disable ON INTERRUPT processing.
DTMFOUT	Produce touch tones on a pin.
EEPROM	Define initial contents of on-chip EEPROM.
ENABLE	Enable ON DEBUG and ON INTERRUPT processing.
ENABLE DEBUG	Enable ON DEBUG processing.
ENABLE INTERRUPT	Enable ON INTERRUPT processing.
END FOR..NEXT	Stop execution and enter low power mode.
FOR..NEXT	Repeatedly execute statements.
FREQOUT	Produce up to two frequencies on a pin.
GOSUB	Call BASIC subroutine at specified label.
GOTO	Continue execution at specified label.
HIGH	Make pin output high.
HSERIN	Hardware asynchronous serial input.
HSEROUT	Hardware asynchronous serial output.
I2CREAD	Read bytes from I2C device.
I2CWRITE	Write bytes to I2C device.
IF..THEN..ELSE..ENDIF	Conditionally execute statements.
INPUT	Make pin an input.
LCDIN	Read from LCD RAM.
LCDOUT	Display characters on LCD.
{LET}	Assign result of an expression to a variable.
LOOKDOWN	Search constant table for value.
LOOKDOWN2	Search constant/variable table for value.
LOOKUP	Fetch constant value from table.
LOOKUP2	Fetch constant/variable value from table.

continued on next page

TABLE 9.2 PicBasic Pro Statement Reference (continued)

STATEMENT	DESCRIPTION
LOW	Make pin output low.
NAP	Power down processor for short period of time.
ON DEBUG	Execute BASIC debug monitor.
ON INTERRUPT	Execute BASIC subroutine on an interrupt.
OUTPUT	Make pin an output.
PAUSE	Delay (1μSec resolution).
PAUSEUS	Delay (1μSec resolution).
PEEK	Read byte from register. (Do not use.)
POKE	Write byte to register. (Do not use.)
POT	Read potentiometer on specified pin.
PULSIN	Measure pulse width on a pin.
PULSOUT	Generate pulse to a pin.
PWM	Output pulse width modulated pulse train to pin.
RANDOM	Generate pseudorandom number.
RCTIME	Measure pulse width on a pin.
READ	Read byte from on-chip EEPROM.
READCODE	Read word from code memory.
RESUME	Continue execution after interrupt handling.
RETURN	Continue at statement following last GOSUB.
REVERSE	Make output pin an input or an input pin an output.
SERIN	Asynchronous serial input (BS1 style).
SERIN2	Asynchronous serial input (BS2 style).
SEROUT	Asynchronous serial output (BS1 style).
SEROUT2	Asynchronous serial output (BS2 style).
SHIFTIN	Synchronous serial input.
SHIFTOUT	Synchronous serial output.
SLEEP	Power down processor for a period of time.
SOUND	Generate tone or white noise on specified pin.
SWAP	Exchange the values of two variables.
TOGGLE	Make pin output and toggle state.
WHILE..WEND	Execute statements while condition is true.

continued on next page

TABLE 9.2 PicBasic Pro Statement Reference (continued)

STATEMENT	DESCRIPTION
WRITE	Write byte to on-chip EEPROM.
WRITECODE	Write word to code memory.
XIN	X-10 input.
XOUT	X-10 output.

Compiling a Program

For operation of the PicBasic Pro Compiler, you'll need a text editor or word processor for creation of your program source file, some sort of PICmicro programmer such as the EPIC Plus Pocket PICmicro Programmer, and the PicBasic Pro Compiler itself. Of course you also need a PC to run it all on.

The sequence of events goes like this:

First create the BASIC source file for the program using your favorite text editor or word processor. If you don't have a favorite, DOS EDIT (included with MS-DOS) or Windows NOTEPAD (included with Windows and Windows 95/98) may be substituted. The source file name should (but isn't required to) end with the extension .BAS. The text file that is created must be pure ASCII text. It must not contain any special codes that might be inserted by word processors for their own purposes. You are usually given the option of saving the file as pure DOS or ASCII text by most word processors.

Program 9.1 provides a good first test for programming a PIC and to test Insectronic's controller board. You can type it in or you can download it from my Web site (http://www.thinkbotics.com/), and follow the links for book software.

The file is named `main-test.bas` and is listed in Program 9.1. The BASIC source file should be created in or moved to the same directory where the PBP.EXE file is located.

PROGRAM 9.1 main-test.bas program listing.

```
'*************************************************************
' Insectronics: build your own six-legged walking robot
' main-test.bas
' Alternatively flash leds and produce random insect noises
' PicBasic Pro Compiler
'**************************************************************
```

```
led_left    var portB.0           ' initialize variables
led_right   var portB.1
piezo       var portB.4
rand        var word
temp1       var byte

start:
        high led_left             ' left LED on
        low led_right             ' right LED off
        gosub randomize
        pause 50

        low led_left              ' left LED off
        high led_right            ' right LED on
        gosub randomize
        pause 50

goto start

randomize:

        Random rand               ' pick a random number
        temp1 = rand & 31 + 64    ' Generate Notes [64..95]
        Sound piezo,[temp1,4]     ' Generate Sound
        return

end
```

Once you are satisfied that the program you have written will work flawlessly, you can execute the PicBasic Pro Compiler by entering `PBP` followed by the name of your text file at a DOS prompt. For example, if the text file you created is named `main-test.bas`, at the DOS command prompt enter:

```
PBP main-test.bas
```

The compiler will display an initialization (copyright) message and process your file. If it likes your file, it will create an assembler source code file (in this case named `MAIN-TEST.ASM`) and automatically invoke its assembler to complete the task. If all goes well, the final PICmicro code file will be created (in this case, `MAIN-TEST.HEX`). If you have made the compiler unhappy, it will issue a string of errors that will need to be corrected in your BASIC source file before you try compilation again.

To help ensure that your original file is flawless, it is best to start by writing and testing a short piece of your program, rather than writing an entire 100,000-line monolith all at once and then trying to debug it from end to end.

If you don't tell it otherwise, the PicBasic Pro Compiler defaults to creating code for the PIC16F84. To compile code for PICmicros other than the 'F84, just use the -P command line option described later in the manual to specify a different target processor. For example, if you intend to run the above program, `main-test.bas`, on a PIC16C74, compile it using the command:

```
PBP -p16c74 main-test.bas
```

The assembler source code listing for `main-test.bas` is shown in Program 9.2 as an example of how much easier it is to program with a compiler like PicBasic Pro. The assembler source code can be used as a guide if you do want to explore assembly language programming because the listing shows the PicBasic statement and the corresponding assembly code on the next line. In the rest of the chapters discussing software, we will not be dealing with assembly code. All we really need to be concerned with is the PicBasic source code and the generated .HEX machine code as listed in Program 9.3.

If you do not have the resources to buy the PicBasic compiler, you can simply type the listings of the .HEX files into a text editor, and save the file with the program name and .HEX extension. All the program listings in the book can also be downloaded from www.thinkbotics.com. However, I do recommend buying a copy of the compiler if you wish to experiment, change, or customize the programs.

PROGRAM 9.2 **Main-test.asm assembler source code.**

```
; PicBasic Pro Compiler 2.40, (c) 1998, 2001 microEngineering Labs,
  Inc. All Rights Reserved.
PM_USED                    EQU   1

  INCLUDE  "16F84A.INC"

RAM_START                  EQU   0000Ch
RAM_END                    EQU   0004Fh
RAM_BANKS                  EQU   00001h
BANK0_START                EQU   0000Ch
BANK0_END                  EQU   0004Fh
EEPROM_START               EQU   02100h
EEPROM_END                 EQU   0213Fh
```

```
; C:\PBP\PBPPIC14.RAM  00012  R0  VAR  WORD BANK0 SYSTEM
  ' System Register
R0        EQU      RAM_START + 000h
; C:\PBP\PBPPIC14.RAM  00013  R1  VAR  WORD BANK0 SYSTEM
  ' System Register
R1        EQU      RAM_START + 002h
; C:\PBP\PBPPIC14.RAM  00014  R2  VAR  WORD BANK0 SYSTEM
  ' System Register
R2        EQU      RAM_START + 004h
; C:\PBP\PBPPIC14.RAM  00015  R3  VAR  WORD BANK0 SYSTEM
  ' System Register
R3        EQU      RAM_START + 006h
; C:\PBP\PBPPIC14.RAM  00016  R4  VAR  WORD BANK0 SYSTEM
  ' System Register
R4        EQU      RAM_START + 008h
; C:\PBP\PBPPIC14.RAM  00017  R5  VAR  WORD BANK0 SYSTEM
  ' System Register
R5        EQU      RAM_START + 00Ah
; C:\PBP\PBPPIC14.RAM  00018  R6  VAR  WORD BANK0 SYSTEM
  ' System Register
R6        EQU      RAM_START + 00Ch
; C:\PBP\PBPPIC14.RAM  00019  R7  VAR  WORD BANK0 SYSTEM
  ' System Register
R7        EQU      RAM_START + 00Eh
; C:\PBP\PBPPIC14.RAM  00020  R8  VAR  WORD BANK0 SYSTEM
  ' System Register
R8        EQU      RAM_START + 010h
; C:\PBP\PBPPIC14.RAM  00026  FLAGS  VAR  BYTE BANK0 SYSTEM
  ' Static flags
FLAGS     EQU      RAM_START + 012h
; C:\PBP\PBPPIC14.RAM  00025  GOP  VAR  BYTE BANK0 SYSTEM
  ' Gen Op Parameter
GOP       EQU      RAM_START + 013h
; C:\PBP\PBPPIC14.RAM  00022  RM1  VAR  BYTE BANK0 SYSTEM
  ' Pin 1 Mask
RM1       EQU      RAM_START + 014h
; C:\PBP\PBPPIC14.RAM  00024  RM2  VAR  BYTE BANK0 SYSTEM
  ' Pin 2 Mask
RM2       EQU      RAM_START + 015h
; C:\PBP\PBPPIC14.RAM  00021  RR1  VAR  BYTE BANK0 SYSTEM
  ' Pin 1 Register
RR1       EQU      RAM_START + 016h
; C:\PBP\PBPPIC14.RAM  00023  RR2  VAR  BYTE BANK0 SYSTEM
  ' Pin 2 Register
RR2       EQU      RAM_START + 017h
; C:\PBP_CODE\MAIN-T~1.BAS  0001  rand  var word
_rand     EQU      RAM_START + 018h
```

```
; C:\PBP_CODE\MAIN-T~1.BAS  00012  temp1  var byte
_temp1       EQU     RAM_START + 01Ah
; C:\PBP\16F84A.BAS  00018  PORTL  VAR  PORTB
_PORTL       EQU     PORTB
; C:\PBP\16F84A.BAS  00019  PORTH  VAR  PORTA
_PORTH       EQU     PORTA
; C:\PBP\16F84A.BAS  00020  TRISL  VAR  TRISB
_TRISL       EQU     TRISB
; C:\PBP\16F84A.BAS  00021  TRISH  VAR  TRISA
_TRISH       EQU     TRISA
#define _led_left         _PORTB_0
#define _led_right        _PORTB_1
#define _piezo            _PORTB_4
#define _PORTB_0          PORTB, 000h
#define _PORTB_1          PORTB, 001h
#define _PORTB_4          PORTB, 004h
  INCLUDE  "MAIN-T~1.MAC"
  INCLUDE  "PBPPIC14.LIB"

; C:\PBP_CODE\MAIN-T~1.BAS   00001   '******************************
; C:\PBP\16F84A.BAS          00012   BANK0   $000C, $004F
; C:\PBP\16F84A.BAS          00013   EEPROM  $2100, $213F
; C:\PBP\16F84A.BAS          00014   LIBRARY "PBPPIC14"

; C:\PBP\16F84A.BAS          00016   include "PIC14EXT.BAS"

; C:\PBP\16F84A.BAS          00023   include "PBPPIC14.RAM"

; C:\PBP_CODE\MAIN-T~1.BAS   00014   start:

  LABEL?L    _start

; C:\PBP_CODE\MAIN-T~1.BAS   00015   high led_left     ' left LED on
  HIGH?T _led_left

; C:\PBP_CODE\MAIN-T~1.BAS   00016   low led_right     ' right LED off
  LOW?T  _led_right

; C:\PBP_CODE\MAIN-T~1.BAS   00017   gosub randomize
  GOSUB?L  _randomize

; C:\PBP_CODE\MAIN-T~1.BAS   00018   pause 50
  PAUSE?C  032h

; C:\PBP_CODE\MAIN-T~1.BAS   00020   low led_left      ' left LED off
  LOW?T  _led_left
```

```
; C:\PBP_CODE\MAIN-T~1.BAS   00021   high led_right    ' right LED on
  HIGH?T _led_right

; C:\PBP_CODE\MAIN-T~1.BAS   00022    gosub randomize
  GOSUB?L  _randomize

; C:\PBP_CODE\MAIN-T~1.BAS   00023    pause 50
  PAUSE?C  032h

; C:\PBP_CODE\MAIN-T~1.BAS   00025    goto start
  GOTO?L _start

; C:\PBP_CODE\MAIN-T~1.BAS   00027    randomize:

  LABEL?L    _randomize

; C:\PBP_CODE\MAIN-T~1.BAS   00029    Random rand     ' pick a random
                                                      ' number
  RANDOM?W _rand

; C:\PBP_CODE\MAIN-T~1.BAS   00030    temp1 = rand & 31 + 64
  ' Generate Notes [64..95]
  AND?WCB  _rand, 05Fh, _temp1

; C:\PBP_CODE\MAIN-T~1.BAS   00031     Sound piezo,[temp1,4]
  ' Generate Sound
  SOUNDPIN?T _piezo
  SOUND?BC _temp1, 004h

; C:\PBP_CODE\MAIN-T~1.BAS   00032     return
  RETURN?

; C:\PBP_CODE\MAIN-T~1.BAS   00034     end
  END?

  END
```

PROGRAM 9.3. **main-test.hex file listing.**

```
:1000000061288F0022088400200928208413 8F088B
:100010003195C28F03091000E0880389000F03011
:100020009103031991000319 8F0303195C28182801
:100030002B2003010C1820088E1F20088E0803199E
:10004000301900F252880060C28262800000F2881
:100050000841780055C280D080C0403198C0A803075
:100060000C1A8D060C198D068C188D060D0D8C0D35
```

```
:100070008D0D5C288F018E00FF308E07031C8F07CB
:10008000031C5C2803308D00DF3048203C288D01A4
:10009000E83E8C008D09FC30031C51288C070318A6
:1000A0004E288C0764008D0F4E280C1857288C1C86
:1000B0005B2800005B280800831303138312640080D
:1000C000080006148316061083128610831686l005
:1000D0008312782032303A200610831606108312DD
:1000E00086148316861083127820323030A206128D5
:1000F00024088C0025088D002B200C08A4000D0876
:10010000A5005F302405A6000630A2001030A00034
:0E01100026088E000430012008006300 8D28B0
:02400E00F53F7C
:00000001FF
```

Using the EPIC Programmer to Program the PIC

There are two steps left—putting your compiled program into the PICmicro micro-controller, and testing it. The PicBasic Pro Compiler generates standard 8-bit Merged Intel HEX (`.HEX`) files that may be used with any PICmicro programmer including the EPIC Plus Pocket PICmicro Programmer shown in Figure 9.4. PICmicros cannot be programmed with BASIC Stamp programming cables.

What follows is an example of how a PICmicro is programmed using the EPIC Programmer with the DOS programming software. If Windows 95/98/NT is available, using the Windows version of EPIC is recommended.

Make sure there are no PICmicros installed in the EPIC Programmer programming socket or any attached adapters. Hook the EPIC Programmer to the PC parallel printer port using a DB25 male-to-DB25 female printer extension cable. Plug the AC adapter into the wall and then into the EPIC Programmer (or attach two fresh 9-volt batteries to the programmer and connect the "Batt ON" jumper). The LED on the EPIC Programmer may be on or off at this point. Do not insert a PICmicro into the programming socket when the LED is on or before the programming software has been started.

Enter:

```
EPIC
```

at the DOS command prompt to start the programming software. The EPIC software should be run from a pure DOS session or from a full-screen DOS session under

FIGURE 9.4 EPIC programmer.

Windows or OS/2. (Running under Windows is discouraged. Windows [all varieties] alters the system timing and plays with the port when you're not looking, which may cause programming errors.)

The EPIC software will take a look around to find where the EPIC Programmer is attached, and get it ready to program a PICmicro. If the EPIC Programmer is not found, check all the above connections and verify that there is not a PICmicro or any adapter connected to the programmer.

Typing:

```
EPIC /?
```

at the DOS command prompt will display a list of available options for the EPIC software.

Once the programming screen is displayed, use the mouse to click on **Open file** or press **Alt-O** on your keyboard. Use the mouse (or keyboard) to select `main-test.hex` or any other file you would like to program into the PICmicro from the dialog box. The file will load and you should see a list of numbers in the window at the left. This is your program in PICmicro code. At the right of the screen there is a display of the configuration information that will be programmed into the PICmicro. Verify that it is correct before proceeding. In general, the Oscillator should be set to XT for a 4-MHz crystal and the Watchdog Timer should be set to ON for PicBasic Pro programs. Most important, Code Protect must be OFF when programming any windowed (JW) PICmicro. You may not be able to erase a windowed PICmicro that has been code protected. Figure 9.5 shows the EPIC MS-DOS interface.

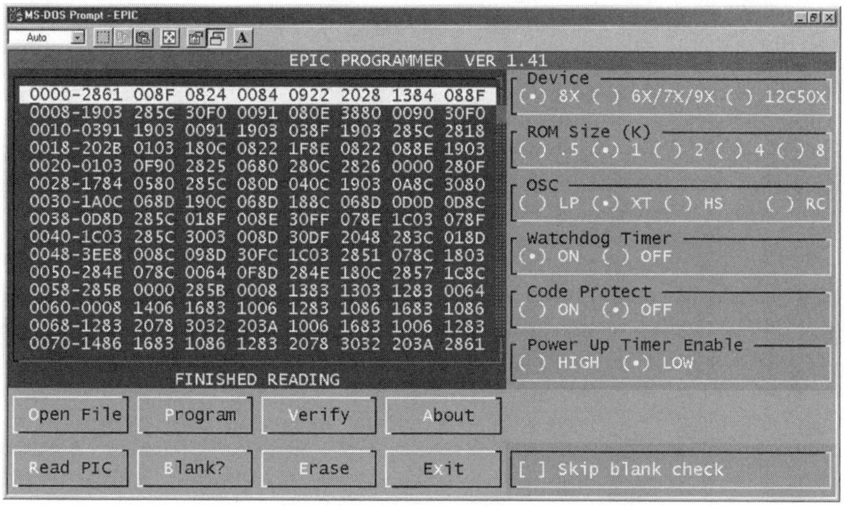

FIGURE 9.5 EPIC graphics user interface.

Insert a PIC 16F84 into the programming socket and click on **Program** or press **Alt-P** on the keyboard. The PICmicro will first be checked to make sure it is blank and then your code will be programmed into it. If the PICmicro is not blank and it is a flash device, you can simply choose to program over it without erasing first. Once the programming is complete and the LED is off, it is time to test your program.

TESTING THE MAIN CONTROLLER BOARD

Now that PIC 16F84 is programmed with the main-test program, it is time to insert the PIC into the I.C. socket on the controller board. Place the PIC into the 18-pin I.C. socket with the notch and pin 1 facing towards the light-emitting diodes using Figure 9.6 as a guide.

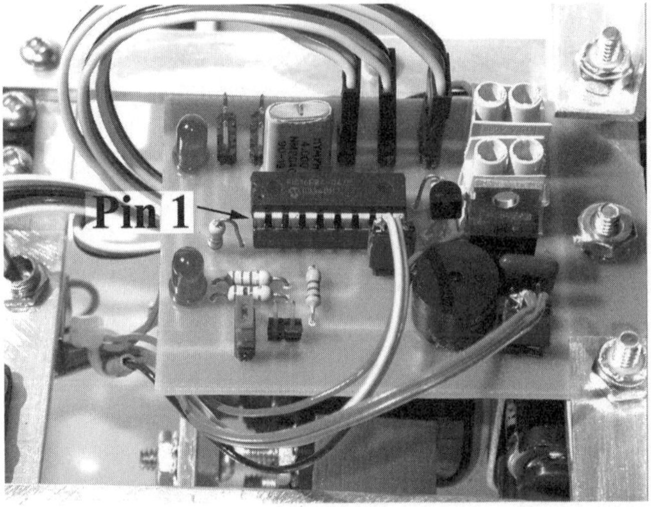

FIGURE 9.6 PIC 16F84 inserted into I.C. socket on controller board.

Make sure that a 9V battery is in the 9V battery holder at the front of the robot and that the 9V battery clip is attached. Turn the switch to the on position towards the back of the robot. If all is well, then Insectronic should be alternatively flashing the left and right light-emitting diodes on and off while making robotic insect noises. This ensures that the 16F84 was programmed and that the controller board is functioning properly.

If nothing is happening when the power is switched on, try going through the process of programming the PIC again and choose the **verify** option from the EPIC user interface. If the chip fails verification, check the RS-232 cable and power supply to the programmer. If that does not work, try using a different 16F84 chip.

If there was not an error when programming the PIC, insert it back into the controller board and make sure that pin 1 is facing toward the light-emitting diodes. Check the battery wiring and verify that the 9V DC polarity is not reversed to the power terminal block. Check the controller board for any missed components or cold solder connections.

Summary

Now that the concept of programming and compiling code for the microcontroller has been covered, we will be able to start having fun with the robot. In the next chapter, Insectronic will evolve through the stages of taking its first steps to the more advanced insect-like exploration behaviors of an artificial life form.

10

EXPERIMENTS

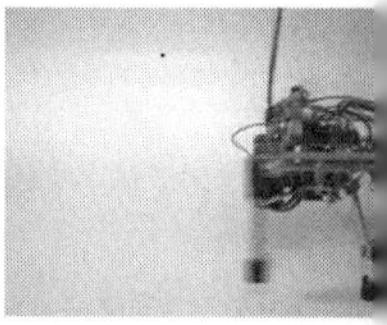

We are now at the point in the book where Insectronic comes to life! Now that all of the hardware is in place, we can start programming insect-like behaviors and experimenting with the robot.

R/C Servo Motors

The R/C servo is a geared direct current motor with a positional feedback control circuit built in as pictured in Figure 10.1. This makes it ideal for use with small robots because the experimenter does not have to worry about motor control electronics.

A *potentiometer* is attached to the shaft of the motor and rotates along with it. For each position of the motor shaft and potentiometer, a unique voltage is produced. The input control signal is a variable-width pulse between 1 and 2 milliseconds (ms) delivered at a frequency between 50 and 60 Hz, which the servo internally converts to a corresponding voltage. The servo feedback circuit constantly compares the potentiometer signal to the input control signal provided by the microcontroller. The internal comparator moves the motor shaft and potentiometer either forward or in reverse until the two signals are the same. Because of the feedback control circuit, the rotor

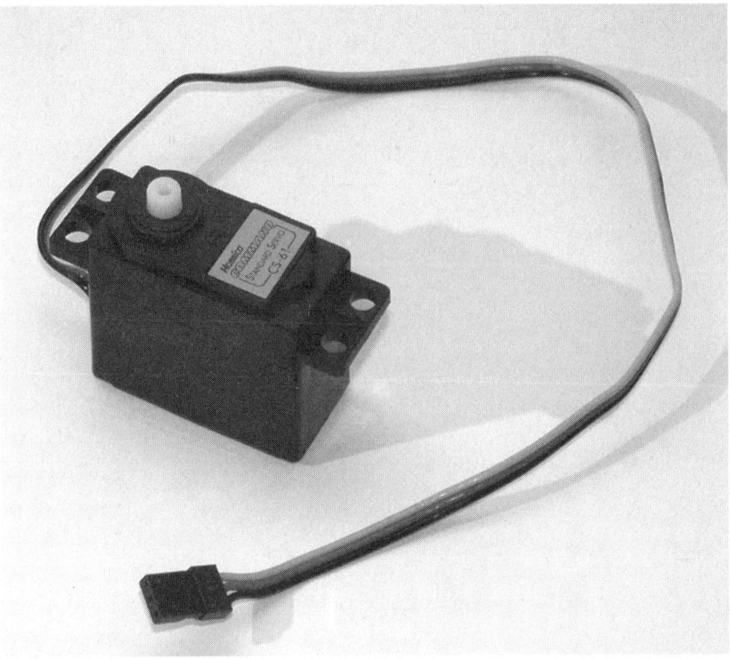

FIGURE 10.1 Standard R/C servo motor.

can be accurately positioned, and will maintain the position as long as the input control signal is applied. The shaft of the motor can be positioned through 180 degrees of rotation depending on the width of the input signal.

The PicBasic Pro language makes servo control with a PIC microcontroller easy using a command called `Pulsout`. The syntax is `Pulsout` *`Pin`*, *`Period`*. A pulse is generated on *`Pin`* of specified *`Period`*. Toggling the pin twice generates the pulse; thus the initial state of the pin determines the polarity of the pulse. *`Pin`* is automatically made an output. *`Pin`* may be a constant, 0-15, or a variable that contains a number between 0 and 15 (e.g., B0) or a pin name (e.g., PORTA.0).

The resolution of `Pulsout` is dependent upon the oscillator frequency. Since we are using a 4-MHz oscillator, the *`Period`* of the generated pulse will be in 10 μs increments. If we want to send a pulse to port B on pin 7 that is 1.4 ms long (at 4 MHz, 10 μs x 140 = 1400 μs or 1.4 ms), the command would be:

```
Pulsout PortB.7,140
```

To illustrate the kind of signal being produced by the microcontroller, see Figure 10.2. The oscilloscope trace for channel 1 was generated with the `Pulsout` command configured to produce a 1.4 ms pulse at 55.68 Hz and the trace for channel 2 is configured for a 6 ms pulse also at 55.68 Hz.

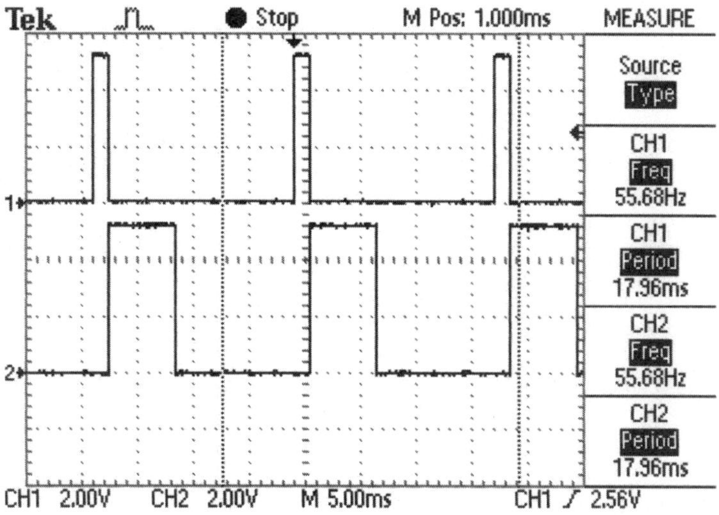

FIGURE 10.2 Oscilloscope display of a 1.4-ms and 6.0-ms pulse train.

For the servos that we are using with Insectronic, a 1.4 ms pulse at a frequency between 50 and 60 Hz will place the servo shaft at approximately middle position. These control pulses from the microcontroller to the servos can be thought of as nerve signals going from an insect's brain to its muscles.

Calibrating the Servos

The next step is to set each of the three servos to their middle positions using software, and then physically adjusting the position of each leg if necessary. We will write a servo control program in PicBasic using the `pulsout` command to do this.

The program is called `servo-cal.bas` and is listed in Program 10.1.

PROGRAM 10.1 servo-cal.bas Program to calibrate Insectronic's servos.

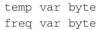

```
'*************************************************************
' servo-cal.bas
' Routine to set servos to their middle positions
' PicBasic Pro Compiler
'*************************************************************

temp var byte                               ' initialize variables
freq var byte
```

```
m_servo var byte
l_servo var byte
r_servo var byte

portb.7 = 0                               ' set initial logic states to 0
portb.6 = 0
portb.5 = 0

freq = 13                                 ' delay value

start:

      m_servo = 140                       ' set servo position values
      l_servo = 140
      r_servo = 140

      gosub servo

goto start


servo:                                    ' servo positioning subroutine
       for temp = 1 to 10
           pulsout portb.7,m_servo
           pulsout portb.6,l_servo
           pulsout portb.5,r_servo
           pause freq                     ' set frequency between 50 - 60 Hz
       next temp
return
```

If you look at the controller board schematic in Figure 10.3, note that the servos are connected to Port B of the microcontroller with the middle servo being on pin 7, the left servo on Pin 6, and the right servo on Pin 5. Our code needs to place a positive square wave pulse with a width of 1.4 microseconds on each of the servo pins at a frequency somewhere between 50 and 60 times every second (Hz) so that the servo will be able to operate properly. For this routine to produce the pulses at approximately 55 Hz, a small delay is introduced in the code after the pulsout commands are called. When the pulsout command is called, it will toggle the initial logic state of the pin for the specified amount of time. To ensure that we produce a positive pulse, we must first set pins 5, 6, and 7 to logic level zeros. The program will be broken down into the main body of code and the servo positioning subroutine. The idea is that we will build onto this program in later experiments and will be able to easily reuse the code. Since the repositioning of servos will be happening quite often, it makes sense to use a subroutine and call it each time it is needed.

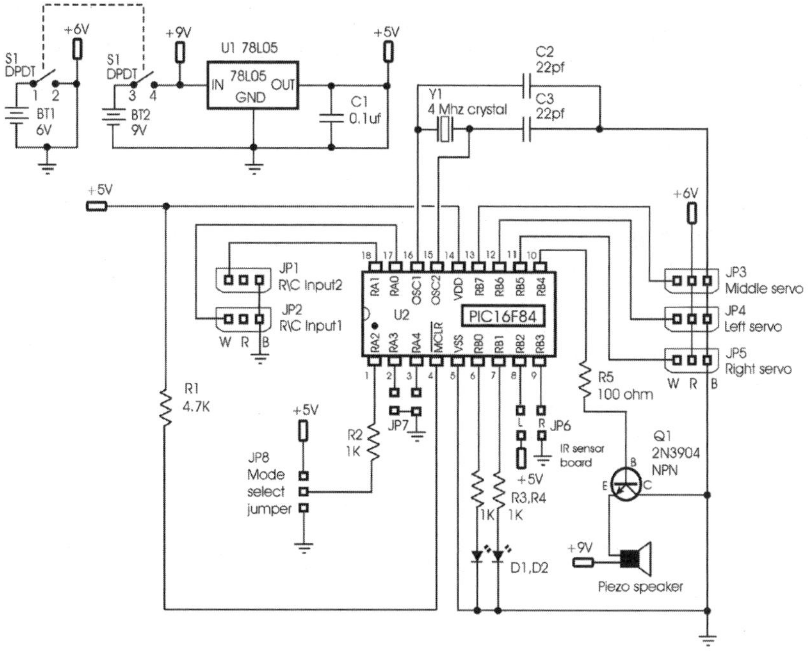

FIGURE 10.3 Main controller board schematic.

Using the instructions in Chapter 9, compile the code, load the corresponding `servo-cal.hex` file into the EPIC programming software, and then program the 16F84 chip. If you do not have the compiler, you can create the `servo-cal.hex` file presented in Program 10.2 by entering it into a text editor and saving it as `servo-cal.hex`.

PROGRAM 10.2 servo-cal.hex file listing.

```
:100000004528A00014200C080D04031940283A20AC
:10001000841320088006640000D280E288C0A03191A
:100020008D0F0B2880064028FF3A84178005402852
:100030008F018E00FF308E07031C8F07031C4028A2
:1000400003308D00DF3026201A288D01E83E8C0019
:100050008D09FC30031C2F288C0703182C288C07D3
:1000600064008D0F2C280C1835288C1C39280000B2
:10007000392808008C098D098C0A03198D0A08009B
:10008000831303138312640008008613061386127 9
:100090000D30A4008C30A6008C30A5008C30A70059
:1000A00052204A280130A80064000B3028020318AF
:1000B000722826088C008D010630840080300120D3
```

```
:1000C00025088C008D01063084004030012027086F
:1000D0008C008D0106308400203001202408182077
:0600E000A80F54280800DF
:02400E00F53F7C
:00000001FF
```

Insert the PIC into the 18-pin I.C. socket on the controller board and turn on the power switch. With the PIC, the internal software program starts execution as soon the chip receives power. All three servos should move to their middle positions and the robot should be positioned as in the diagram in Figure 10.4.

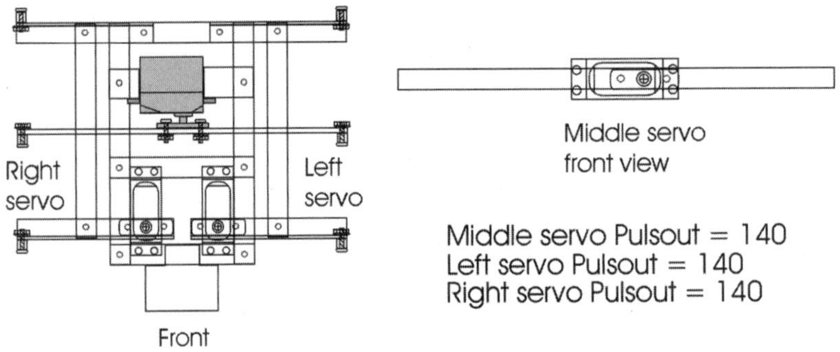

FIGURE 10.4 **Pulsout values for Insectronic's middle leg positions.**

If the legs are not at right angles, then remove the servo horn screws while the robot's power is on. For each servo, pull the servo horn off of the servo shaft and reposition each leg so that they are the same as the diagram in Figure 10.4 where pulsout for each servo = 140. Once that is done, we are now ready to calibrate Insectronic's infrared sensors.

Calibrating the Infrared Sensors

The program called `infrared-cal.bas`, listed in Program 10.3, will be used to fine tune the infrared sensor board. With the circuit operating at a frequency of approximately 40.9 kHz, the I.R. modules will be constantly placing a logic level 1 (+5 vdc) on the output pin. When a modulated infrared signal is reflected back, the module will place a logic level 0 (0 volts) on the output pin. We will use the microcontroller to monitor the left and right infrared modules' output pins. When a logic level of 0 is received, it will turn on the corresponding LED on the controller board so that we can visually see what is going on. Compile `infrared-cal.bas` and then program the PIC 16F84 with `infrared-cal.hex` listed in Program 10.4 using the procedure in Chapter 9. When the programming is finished, insert the PIC into the I.C. socket on the controller board and turn the power on.

PROGRAM 10.3 infrared-cal.bas program listing.

```
'**********************************************************
' infrared-cal.bas
' Routine to calibrate the infrared sensor board
' PicBasic Pro Compiler
'**********************************************************

trisa = %11111111                ' set porta to inputs
trisb = %00001100                ' set portb pins 2 & 3 to inputs


start:

      low portb.0

      If portb.3 = 0 then       ' Check right i.r. module output for
                                ' logic 0
          high portb.0          ' turn on right light emitting diode
      else
          low portb.0
      endif

      low portb.1

      If portb.2 = 0 then        ' Check left i.r. module output for
                                 ' logic 0
          high portb.1           ' turn on left light emitting diode
      else
          low portb.1
      endif

goto start

end
```

PROGRAM 10.4 infrared-cal.hex file listing.

```
:1000000001288316FF3085000C308600831206100D
:1000100083160610640083128619132806148316AB
:10002000061083121728061083160610831286100F6
:1000300083168610640083120619232886148316FB
:1000400086108312272886108316861083120728AD
:02400E00F53F7C
:00000001FF
```

Place the robot on a table facing towards you with no solid objects surrounding it. The left and right light-emitting diodes on the controller board should be off. Move your hand towards the left emitter and detector module. At a distance of 6 inches, the left LED should start to flicker, and at 5 inches should turn on solid. Repeat this procedure for the right emitter and detector module. Next, move your hand towards the robot directly in front of the detector board. At a distance of 4 inches, both the left and right LEDs on the controller board should start to flicker and at 3 inches should turn on solid.

If one or both of the LEDs on the controller board are on or flickering when no solid objects are close to the robot, then the frequency of the I.R. sensor board will need to be adjusted. Facing towards the front of the robot, turn potentiometer R3 in very small increments counter clockwise until both LEDs are on solidly. Then slowly rotate the potentiometer clockwise until both LEDs flicker and then turn off completely. Figure 10.5 shows potentiometer R3 being adjusted. If this does not work, then use the instructions in Chapter 7 to set the frequency to approximately 40.9 kHz and then try this procedure again if necessary.

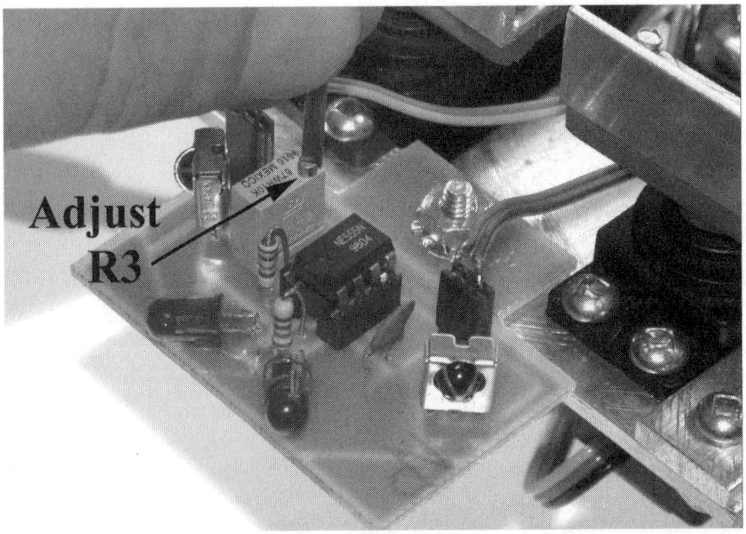

FIGURE 10.5 Adjusting potentiometer R3 to find the correct frequency.

Walking

In order to make Insectronic walk, we will need to develop a walking-gait routine that will coordinate the leg movements in the proper sequence through time and space. As discussed in Chapter 2, Insectronic will be programmed to use a modified tripod gait to achieve locomotion. The primary leg positions and the corresponding pulsout values needed for programming the walking movements are shown in Figure 10.6. Let's say that we want to move the right servo from the position shown in the bottom diagram of Figure 10.6 with a pulsout value of 160 (1.6 ms pulse width) to the position in the top diagram of Figure 10.6 with a pulsout value of 120 (1.2 ms pulse width). When the servo is given the 1.2-ms pulse train, it will move through the entire range of movement between the position that the servo is currently in, and the position that needs to be reached. The great thing about using a servo is that we do not need to program a software routine to cover the movement of all the position points in between. This is taken care of by the servo's internal electronics. All we need to do is set the pulse width corresponding to the position that is required, and then the servo does the rest.

To program the walking sequence, we will start with the assumption that each of the robot's legs is in the middle position as in Figure 10.4. As you will see later, the starting position of the legs is not really important because after three moves, all the legs will be in sync.

Back in Chapter 2, walking gaits were discussed in detail, as well as the specific sequence that would be required to produce locomotion with a three-servo hexapod robot. Now that the pulse-width values have been determined for the primary leg positions, we can use these values and the sequencing information to program the leg movements so that the robot will walk forward. To make the robot walk backwards, it is just a matter of running this sequence in reverse for the left and right servos, while the positions for the middle servo remain the same.

To make the discussion easier, the legs of the robot have been numbered from 1 to 6 as shown in frame 1 of Figure 10.7. The servo positioned in the middle of the robot's body is attached to legs 2 and 5. It is used to rock the body back and forth and in turn lifts up legs 1, 3, and 5 or legs 2, 4, and 6. Legs 1 and 3 are controlled by a single servo (which has been referred to as the robot's left servo) and move together via a mechanical linkage. Legs 4 and 6 are also controlled by a single servo (the robot's right) and move together via a mechanical linkage.

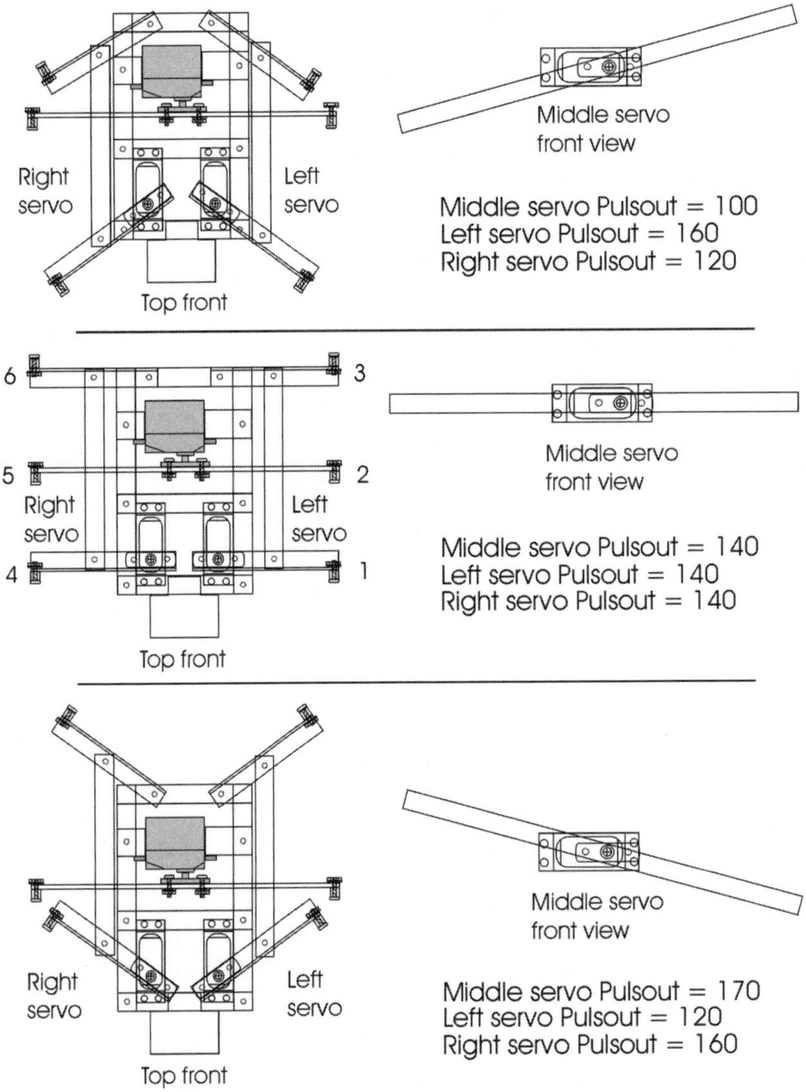

FIGURE 10.6 Primary leg positions and the corresponding pulsout values.

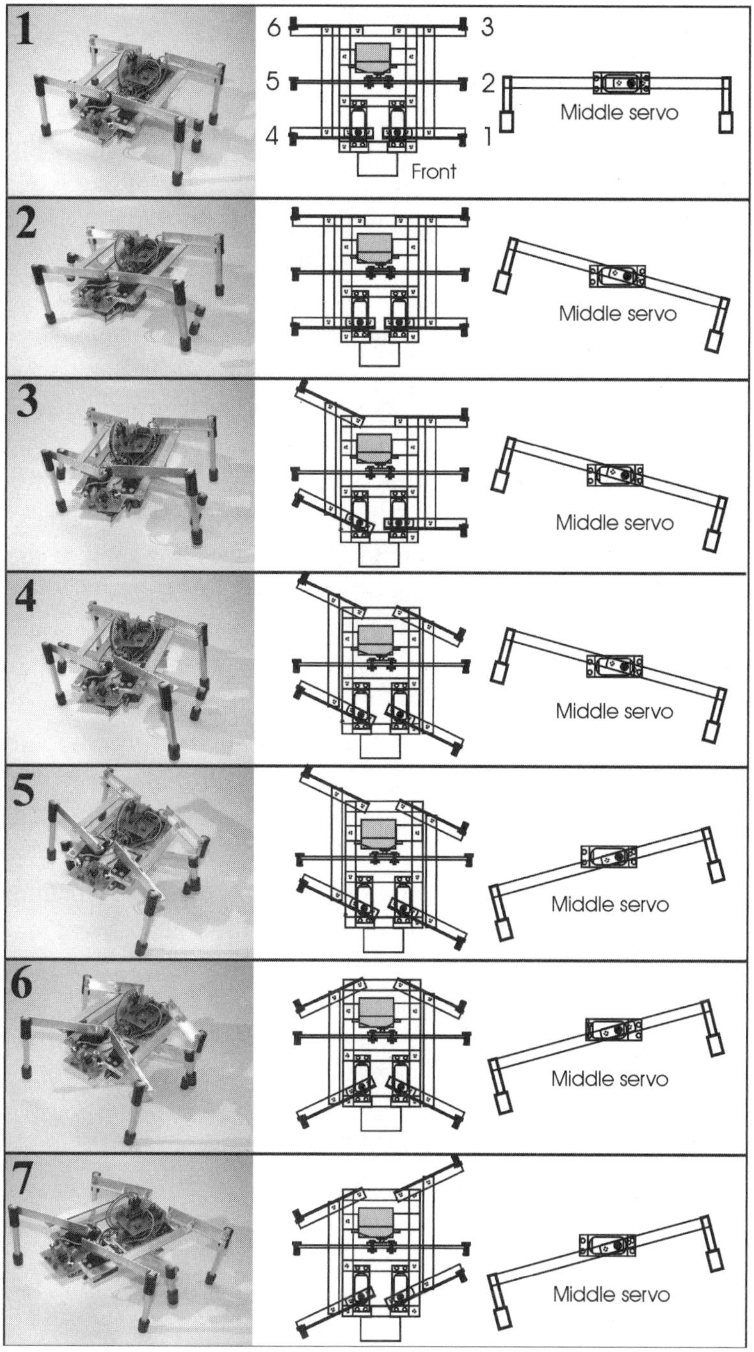

FIGURE 10.7 Leg position sequence to achieve forward walking.

FORWARD AND REVERSE WALKING SEQUENCE

To start the forward walking sequence, all legs are flat on the ground with the servos in their middle positions as shown in frame 1 of Figure 10.7. The first move shown in frame 2 of Figure 10.7 is to move leg 2 down, which lifts legs 1, 3, and 5 up off the ground. In frame 3, leg 2 remains down and is used as a swivel as legs 4 and 6 are moved backwards, propelling the right side of the robot forward. In frame 4, leg 2 remains down while legs 1 and 3 are moved forward in anticipation of the next move. In frame 5, leg 5 is now moved down, lifting legs 2, 4, and 6 up off the ground. In frame 6, leg 5 remains down and legs 4 and 6 are moved forward in anticipation of the next move. In frame 7, leg 5 remains down and acts as a swivel as legs 1 and 3 are moved backwards, propelling the left side of the robot forward. For the robot to continue walking forward, the sequence is repeated from frames 2 to 7.

To simplify the sequence for the purposes of programming and to speed up the walking gait, the steps in which legs are moved forward in anticipation of the next move can be performed at the same time that the opposite set of legs is propelling the robot forward. In this case, the move in frame 3 can be performed at the same time as the move in frame 4. Also, the moves in frames 6 and 7 can be performed at the same time. This means that the software will only have to set the servo positions four times to complete one forward walking sequence, as listed in Table 10.1. The pulsout values that we will be using to program the robot are taken from those listed in Figure 10.6 and correspond to the positions in Figure 10.7. For the robot to walk in reverse, the above sequence is run in reverse for the left and right servo legs while the middle servo values remain the same as discussed.

Tables 10.1 and 10.2 show the exact four-step sequences and the corresponding servo positions that will be needed when programming the robot to walk forward and reverse.

TABLE 10.1 Forward Walking Sequence with Pulsout Leg Position Values

SEQUENCE NUMBER	MIDDLE SERVO	LEFT SERVO	RIGHT SERVO
1	170	—	—
2	170	160	160
3	100	—	—
3	100	120	120

TABLE 10.2 Reverse Walking Sequence with Pulsout Leg Position Values

Sequence Number	Middle Servo	Left Servo	Right Servo
1	170	—	—
2	170	120	120
3	100	—	—
3	100	160	160

ROTATING LEFT AND RIGHT WALKING SEQUENCE

To discuss how the robot accomplishes turning left and right, we will start with all of the robot's legs flat on the ground, as shown in frame 1 of Figure 10.8.

The robot's first action, shown in frame 2 of Figure 10.8, is to move leg number 2 down, which lifts legs 1, 3, and 5 off of the ground. In frame 3 of Figure 10.8, leg 2 remains in the down position, acting as a pivot point while legs 4 and 6 are moved backwards, swiveling the robot to the left. At the same time, legs 1 and 3 (which are lifted off the ground) are moved backwards in anticipation of the next move. In frame 4, the middle servo moves leg 5 to the downward position, lifting legs 2, 4, and 6 off of the ground. In frame 5, leg 5 remains in the down position while legs 1 and 3 move forward, also causing the robot to pivot to the left. At the same time, legs 4 and 6 are moved forward in anticipation of the next move. Because the middle leg is used as a swivel, the robot can easily turn on the spot. For the robot to turn to the right, the sequence is reversed for the left and right servos, while the middle servo values stay the same as listed.

Tables 10.3 and 10.4 show the exact four-step sequences and the corresponding servo positions that will be needed when programming the robot to turn left and right.

TABLE 10.3 Left-Turn Walking Sequence with Pulsout Leg Position Values

Sequence Number	Middle Servo	Left Servo	Right Servo
1	170	—	—
2	170	120	160
3	100	—	—
3	100	160	120

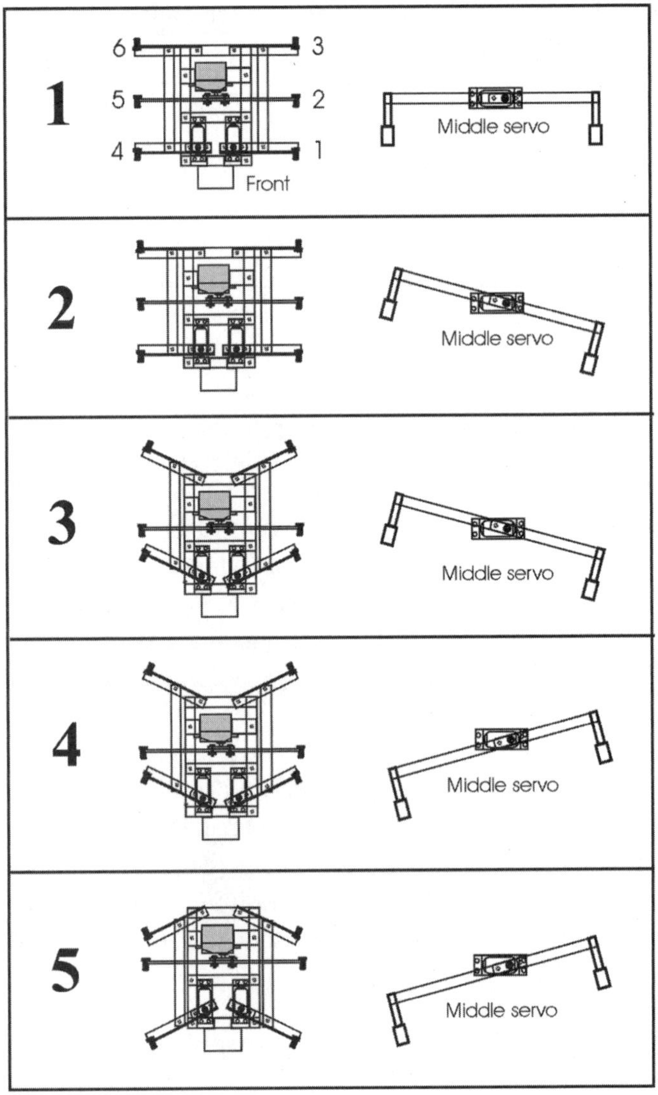

FIGURE 10.8 Robot leg positions for turning sequence.

TABLE 10.4 Right-Turn Walking Sequence with Pulsout Leg Position Values

Sequence Number	Middle Servo	Left Servo	Right Servo
1	170	—	—
2	170	160	120
3	100	—	—
3	100	120	160

Now we will take the information about sequencing the leg movements and put it together in a program to demonstrate the four basic walking gaits. Program 10.5, which follows, is broken down into four parts, and will coordinate the robot's legs to move the robot in a forward walking motion, turn the robot 90 degrees to the left, walk the robot in reverse, turn the robot to the right, and then repeat the entire sequence. The program operates by setting three variables with the pulsewidth values that we want for each servo position as listed in Tables 10.1, 10.2, 10.3, and 10.4, then calling a subroutine to set the servos. The subroutine uses the pulsout command discussed earlier to send the proper pulse width value to each servo according to the values contained in the variables. Compile the program and then start the EPIC programming software. Load the `walking-gaits.hex` file and program the 16F84 PIC. The `walking-gaits.hex` file is listed in program 10.6 if you do not have the compiler. Type it into a text file and save it as `walking-gaits.hex` or download the file from http://www.thinkbotics.com/.

PROGRAM 10.5 walking-gaits.bas program listing.

```
'***********************************************************
' walking-gaits.bas
' walking forward, reverse, turning left and turning right
' PicBasic Pro Compiler - microEngineering Labs
'***********************************************************

temp var byte              ' initialize variables
timer var byte
m_servo var byte
l_servo var byte
r_servo var byte

portb.7 = 0                ' set servo channels to start low
portb.6 = 0
portb.5 = 0

m_servo = 140              ' value for servos in middle position
l_servo = 140
r_servo = 140

walk_forward:

              for temp = 1 to 7
                  m_servo = 170
                  gosub servo
                  l_servo = 160
```

```
                    r_servo = 160
                    gosub servo
                    m_servo = 100
                    gosub servo
                    l_servo = 120
                    r_servo = 120
                    gosub servo
               next temp

turn_left:

               for temp = 1 to 3
                    m_servo = 170
                    gosub servo
                    l_servo = 120
                    r_servo = 160
                    gosub servo
                    m_servo = 100
                    gosub servo
                    l_servo = 160
                    r_servo = 120
                    gosub servo
               next temp

walk_reverse:
               for temp = 1 to 7
                    m_servo = 170
                    gosub servo
                    l_servo = 120
                    r_servo = 120
                    gosub servo
                    m_servo = 100
                    gosub servo
                    l_servo = 160
                    r_servo = 160
                    gosub servo
               next temp

turn_right:

               for temp = 1 to 3
                    m_servo = 170
                    gosub servo
                    l_servo = 160
                    r_servo = 120
                    gosub servo
                    m_servo = 100
```

```
                    gosub servo
                    l_servo = 120
                    r_servo = 160
                    gosub servo
                next temp

goto walk_forward

servo:                                         ' subroutine to set servos

        for timer = 1 to 10
        pulsout portb.7,m_servo
        pulsout portb.6,l_servo
        pulsout portb.5,r_servo
        pause 13
        next timer
return

end
```

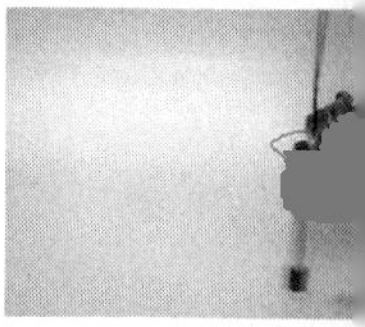

PROGRAM 10.6 **walking-gaits.hex file listing.**

```
:100000004528A00014200C080D04031940283A20AC
:1000100084132008800664000D280E288C0A03191A
:100020008D0F0B2880064028FF3A84178005402852
:100030008F018E00FF308E07031C8F07031C4028A2
:1000400003308D00DF3026201A288D01E83E8C0019
:100050008D09FC30031C2F288C0703182C288C07D3
:1000600064008D0F2C280C1835288C1C39280000B2
:10007000392808008C098D098C0A03198D0A08009B
:100080008313031383126400080086130613861279
:100090008C30A5008C30A4008C30A6000130A70065
:1000A00064000830270203186728AA30A500B3208F
:1000B000A030A400A030A600B3206430A500B32077
:1000C0007830A4007830A600B320A70F5028013064
:1000D000A70064000430270203188028AA30A50076
:1000E000B3207830A400A030A600B3206430A5006F
:1000F000B320A030A4007830A600B320A70F692851
:100100000130A70064000830270203189928AA309C
:10011000A500B3207830A4007830A600B320643066
:10012000A500B320A030A400A030A600B320A70FE4
:1001300082280130A700640004302702031 8B22887
:10014000AA30A500B320A030A4007830A600B320C8
:100150006430A500B3207830A400A030A600B320FE
:10016000A70F9B284E280130A80064000B302802FE
:100170000318D32825088C008D01063084008030B8
```

```
:1001800001202408 8C008D01063084004030 0120BD
```

Infrared Module Software Analysis

At this point, Insectronic has the capability to walk. What we need to do now is take input from the infrared sensors so that the robot can change its behavior to safely avoid any obstacles it may encounter while walking through its environment. A software subroutine will need to be developed to monitor the infrared sensor modules, perform signal processing to clean up any background noise or transient signals to make the information more useful, and then return results to the robot's main program. In this behavior-based method of artificial intelligence, the robot will continue on with the dominant behavior of exploring, and will change that course of action immediately based on sensor input.

Program 10.7 is called `ir-analysis.bas` and will be used to develop the infrared sensor module subroutine. What we want the main program to do is to call the subroutine and have the subroutine simply return two values, one for each sensor module, after performing the signal processing. The value returned will be either a 1 or a 0, with 0 indicating that no object was sensed and 1 indicating that an object is present. These values will be stored in variables `ir_left` and `ir_right`. When the program execution is returned back to the main program, certain decisions can easily be made based on this information. In this experiment, we will be turning on the corresponding LED for each sensor and producing an insect-like sound to acknowledge that an object has been sensed.

The infrared subroutine takes 30 samples from each module and counts the number of positive hits received from each module. Because of stray infrared and signals from the environment, the modules are constantly producing false positive signals that are referred to as "noise." The average acceptable amount of noise that is picked up by the sensor modules is called the *noise floor*. What the routine needs to do is to set a threshold point above the typical amount of noise and report a sensed object only if the number of positive signals received throughout the number of samples taken exceeds the noise floor. With the PNA4602M sensor modules, I found that the typical false positive was actually very low—4 for every 30 samples taken. To be on the safe side, the threshold is set at 15 for every 30 samples to ensure that an object is present. By changing the threshold values, you can change the sensitivity and distance detection response of the modules. If you want a more accurate reading, the `num_samples` value can be increased but will take more time for the routine to execute. Compile `ir-analysis.bas` listed in Program 10.7 and then program the 16F84 PIC with the `ir-analysis.hex` file shown in Program 10.8. When you run the program you will notice that there is now a clearly defined point when the LEDs turn on solidly, not as in the calibration program where the LEDs started out with a few flickers, got brighter and then turned on solid. This will make it easier to program the main routine now that we have a threshold set.

PROGRAM 10.7 ir-analysis.bas program listing

```
'*****************************************************
' ir_analysis.bas
' Program to analize output from infrared sensor modules
' PicBasic Pro Compiler
'*****************************************************

trisa = %11111111                ' set porta to inputs
trisb = %00001100                ' set portb pins 2 & 3 to inputs

temp          var   byte         ' initialize variables
ir_left       var   byte
ir_right      var   byte
l_count       var   byte
r_count       var   byte
l_threshold   var   byte
r_threshold   var   byte
num_samples   var   byte

num_samples = 30           ' number of samples taken per i.r. module
l_threshold = 15           ' noise floor threshold for left detector
r_threshold = 15           ' noise floor threshold for right detector

start:

      gosub infrared        ' get sensor values from infrared
                            ' subroutine

      if ir_left = 1 then   ' if left detector senses an object then
          high portb.1        ' turn on left LED
          sound portb.4,[130,1,90,2,100,1,110,2]   ' output left
                                                   ' insect noise
      else
          low portb.1
```

```
     endif

     if ir_right = 1 then       ' if right detector senses an object
  then
         high port              ' turn on right LED
         sound portb.4,[140,1,120,2,110,1,100,2]    ' output right
                                                    ' insect noise

     else
         low portb.0
     endif

goto start

infrared:

l_count = 0
r_count = 0
ir_left = 0
ir_right = 0

         for temp = 1 to num_samples    ' take 30 samples per
                                        ' detector

             if portb.2 = 0 then        ' if left module senses an
                                        ' object
                l_count = l_count + 1   ' increase the left count
             endif

             if portb.3 = 0 then         ' if right module senses an
                                         ' object
                r_count = r_count + 1   ' increase the right count
             endif

         next

         if l_count >= l_threshold then  ' if left count is greater
                                         ' than the threshold
            ir_left = 1                  ' then an object was
                                         ' sensed with left
                                         ' detector
         endif

         if r_count >= r_threshold then   ' if right count is
                                          ' greater than the
                                          ' threshold
            ir_right = 1                  ' then an object was
                                          ' sensed with right
```

```
                                              ' detector
          endif

return                                        ' return to main program

end
```

PROGRAM 10.8   ir-analysis.hex file listing

```
:100000003F288F0022088400200928208413 8F08AD
:1000100003193A28F03091000E0880389000F03033
:100020009103031991000319 8F0303193A28182823
:100030002B2003010C1820088E1F20088E0803199E
:100040000301900F252880060C28262800000F2881
:1000500084178 0053A280D080C0403198C0A803097
:100060000C1A8D060C198D068C188D060D0D8C0D35
:100070008D0D3A28831303138312640080083163E
:10008000FF3085000C30860083121E30A8000F3030
:10009000A7000F30AA00912064002408013C031D32
:1000A0006A28861483168610063083 12A200103048
:1000B000A00082308E00013001205A308E000230C4
:1000C0000012064308E00013001206E308E0002303D
:1000D0000012 06E2886108316861083 12640025087E
:1000E00013C031D8C280614831606100630831 26B
:1000F000A2001030A0008C308E000130012078303A
:100100008E00023001206E308E0001300120 6430FC
:100110008E00023001209028061083 1606108312EC
:100120004B28A601A901A401A501013 0AB00640080
:100130002B082802031CA62864000619A028A60A7A
:10014 00064008619A428A90AAB0F97286400270821
:100150002602031CAD280130A40064002A082902ED
:0E016000031CB4280130A50008006300B52878
:02400E00F53F7C
:00000001FF
```

Behavior-based Subsumption Architecture

Behavior-based control is a modern robot control technique that was inspired by insects. Insects have very small brain capacities, and yet they manage to find food, navigate over terrain without losing their bearings, escape from predators, work together for one common goal, and react to all sorts of adverse conditions. By the 1980s robot research had come to the conclusion that insects were still more intelligent than the robots they had been creating. The problem with using standard mathematical programming techniques was that the robots could not handle unstructured environments without becoming completely overwhelmed. It was impossible mathematically to describe the robot's environment and then have it react in any reasonable amount of time.

To solve these problems, the first behavior-based robot programming scheme for artificial intelligence was introduced by Rodney Brooks in what he called "subsumption architecture." What he proposed was to do away with the complex mathematical control algorithms in favor of reactive responses the same way that insects manage to navigate through the world.

A behavior-based system is made up of a number of simple, individual behaviors. Each behavior has a task to perform, but no single behavior by itself provides all the control that the robot needs to accomplish a certain job. For example, the robot could start out with an obstacle-avoidance behavior, where it would stop and turn when an object was encountered. On top of this is added the capacity to avoid moving objects. Next the ability to scan the room, locate an object, and then walk towards the object is added. At the same time, the obstacle-avoidance abilities remain intact. On top of this is added the exploration behavior. Nothing is learned, and yet the robot now seems to have goal-directed behavior much like that of an insect. The big debate among AI researchers is that while using this scheme, the robot has no internal representation of the external world. The good thing about this is that these robots can be easily programmed and debugged, they require little code space, and with few internal states they do not need a large amount of RAM. The elegance and power of behavior-based systems comes from combining these individual behaviors in such a way that something complex, interesting, and often surprising emerges.

EXPLORATION WITH OBSTACLE AVOIDANCE

We will now take the code from the individual programming experiments and put it together into an entire artificial intelligence control program for Insectronic. The robot has been providing us with feedback via the two light-emitting diodes and with sound through the piezo speaker on the controller board. The robot now has the ability to walk in all directions and to sense objects. Table 10.5 is a truth table of the possible values received from the infrared subroutine and will be used to directly determine the robot's actions.

TABLE 10.5 Object Location Based on Infrared Sensor Output

IR_LEFT	IR_RIGHT	LOCATION OF OBJECT	BEHAVIOR = IR_LEFT * 2 + IR_RIGHT
0	0	None detected	0
0	1	To the right	1
1	0	To the left	2
1	1	Directly in front	3

If the two variables are added together after multiplying ir_left * 2, we can use this value to determine the robot's course of action as shown in Table 10.6.

TABLE 10.6 Robot Behavior Based on Infrared Sensor Values

BEHAVIOR VALUE	COURSE OF ACTION
0	Continue to explore the environment
1	Stop, turn to the left and continue to explore
2	Stop, turn to the right and continue exploring
3	Stop, back up, randomly turn left or right

To coordinate the proper actions based on the sensor values, we can use a command called `BRANCH` with the following syntax:

```
BRANCH Index, [Label{,Label...}]
```

`BRANCH` causes the program to jump to a different location based on a variable index. This is similar to On..Goto in other BASIC programs. *`Index`* selects one of a list of labels. Execution resumes at the indexed label. For example, if *`Index`* is zero, the program jumps to the first label specified in the list, if *`Index`* is one, the program jumps to the second label, and so on. If *`Index`* is greater than or equal to the number of labels, no action is taken and execution continues with the statement following the `BRANCH`. Up to 255 (256 for 18Cxxx) *`labels`* may be used in a `BRANCH`. *`Label`* must be in the same code page as the `BRANCH` *`Index`* instruction.

For our robotic control program, the index will be the appropriate robot behavior as determined by the variable behavior:

```
BRANCH behavior,[walk_forward, turn_left, turn_right, walk_reverse]

' Same as:
' If  behavior = 0 Then walk_forward (goto walk_forward)
' If  behavior = 1 Then turn_left (goto turn_left)
' If  behavior = 2 Then turn_right (goto turn_right)
' If  behavior = 3 Then walk_reverse (goto walk_reverse)
```

We also need to add the random turn direction when the robot encounters an object directly to the front. We will use a variable named *decision* to hold the random number. This can be accomplished by using the `RANDOM` command with the following syntax:

```
RANDOM Var
```

This command will perform one iteration of pseudorandomization on *`Var`*. *`Var`* should be a 16-bit variable. Array variables with a variable index may not be used in `RANDOM` although array variables with a constant index are allowed. *`Var`* is used both as the seed and to store the result. The pseudorandom algorithm used has a walking length of 65535 (only zero is not produced).

In our program, the statement will look like this:

```
RANDOM decision              ' create a random number and place it in
                             ' variable decision
```

The value of bit 0 is accessed by specifying the bit we want:

```
decision.0
```

The program listing for `insect-explore.bas` is shown in Program 10.8. You will notice that when the program starts, a delay is produced by flashing the LEDs and producing sounds. This delay lasts for about 1 second, and the purpose is to let the infrared modules stabilize before the robot starts the walking routine. Use the `insect-explore.hex` file to program the 16F84 PIC, listed in Program 10.10.

PROGRAM 10.9 insect-explore.bas program listing.

```
'*************************************************************
' insect-explore.bas
' exploration with obastacle avoidance control program
' PicBasic Pro Compiler
'*************************************************************
```

```
trisa = %11111111                    ' set porta to inputs
trisb = %00001100                    ' set portb pins 2 & 3 to inputs

m_servo  var byte                    ' initialize variables
l_servo  var byte
r_servo  var byte
temp     var byte
timer    var byte
decision var word
ir_left  var byte
ir_right var byte
l_count  var byte
r_count  var byte
l_threshold var byte
r_threshold var byte
num_samples var byte
behavior var byte

num_samples = 30          ' number of samples taken per i.r. module
l_threshold = 5           ' noise floor threshold for left detector
r_threshold = 5           ' noise floor threshold for right detector

for temp = 1 to 10                          ' delay for approximately 1
                                            ' second by
    sound portb.4,[90 + (temp*3),2]         ' flashing LED's and making
                                            ' sounds so that
    toggle portb.1                          ' the infrared detecor board
                                            ' has time to
    toggle portb.0                          ' stabalize on power up.
    pause 100
next temp

start:

      gosub infrared          ' get sensor values from infrared
                              ' subroutine

      portb.1 = ir_left       ' turn on LED if an object was detected
      portb.0 = ir_right

      behavior = ir_left * 2 + ir_right

      branch behavior,[walk_forward, turn_left, turn_right,
  walk_reverse]

walk_reverse:
```

```
            sound portb.4,[90,1,80,2,125,1,90,2,100,2]

            for temp = 1 to 2
                m_servo = 170
                gosub servo
                l_servo = 120
                r_servo = 120
                gosub servo
                m_servo = 100
                gosub servo
                l_servo = 160
                r_servo = 160
                gosub servo
            next temp

            random decision

            if decision.0 = 1 then turn_right

turn_left:

            sound portb.4,[140,1,80,2,125,1,95,2]

            for temp = 1 to 2
                m_servo = 170
                gosub servo
                l_servo = 120
                r_servo = 160
                gosub servo
                m_servo = 100
                gosub servo
                l_servo = 160
                r_servo = 120
                gosub servo
            next temp
            goto start

turn_right:

            sound portb.4,[140,1,120,2,110,1,100,2]

            for temp = 1 to 2
                m_servo = 170
                gosub servo
                l_servo = 160
                r_servo = 120
                gosub servo
```

```
                m_servo = 100
                gosub servo
                l_servo = 120
                r_servo = 160
                gosub servo
            next temp
            goto start

walk_forward:

                m_servo = 170
                gosub servo
                l_servo = 160
                r_servo = 160
                gosub servo
                m_servo = 100
                gosub servo
                l_servo = 120
                r_servo = 120
                gosub servo
goto start

infrared:

l_count = 0
r_count = 0
ir_left = 0
ir_right = 0

         for temp = 1 to num_samples    ' take 30 samples per
                                        ' detector

             if portb.2 = 0 then        ' if left module senses an
                                        ' object
                l_count = l_count + 1   ' increase the left count
             endif

             if portb.3 = 0 then        ' if right module senses an
                                        ' object
                r_count = r_count + 1   ' increase the right count
             endif

         next

         if l_count >= l_threshold then    ' if left count is
                                           ' greater than the
                                           ' threshold
```

```
            ir_left = 1                     ' then an object was
                                            ' sensed with left
                                            ' detector
        endif

        if r_count >= r_threshold then      ' if right count is
                                            ' greater than the
                                            ' threshold
            ir_right = 1                    ' then an object was
                                            ' sensed with right
                                            ' detector
        endif

return                                      ' return to main program

servo:                                      ' subroutine to set servos

       for timer = 1 to 10
       pulsout portb.7,m_servo
       pulsout portb.6,l_servo
       pulsout portb.5,r_servo
       pause 13
       next timer
return

end
```

PROGRAM 10.10 **insect-explore.hex file listing.**

```
:100000009128A2003B200C080D0403198C2886209F
:1000100084132208800664000D280E288C0A031918
:100020008D0F0B2880068C288F002408840022095D
:100030003C2084138F0803198C28F03091000E089F
:1000400080389000F03091030319910003198F0359
:1000500003198C282B283F2003010C1822088E1F1F
:1000600022088E0803190301900F382880061F28E4
:1000700039280000222 8FF3A841780058C280D08B3
:100080000C0403198C0A80300C1A8D060C198D068D
:100090008C188D060D0D8C0D8D0D8C288F018E000A
:1000A000FF308E07031C8F07031C8C2803308D0044
:1000B000DF305C2050288D01E83E8C008D09FC303B
:1000C00031C65288C07031862288C0764008D0FB9
:1000D00062280C186B288C1C6F2800006F28080001
:1000E000103094008D018C01930C920C031C7E281F
:1000F0000E088C070F0803180F0F8D078D0C8C0C42
```

```
:10010000910C900C940B742810088C288C098D0984
:100110008C0A03198D0A08008313031383126400E9
:1001200008008316FF3085000C30860083121E30D5
:10013000AF000530AD000530B2000130B3006400FF
:100140000B3033020318C8280630A4001030A20078
:1001500033089200930103308E008F0170209E00BF
:1001600011089F005A309E0703189F0A1E088E0030
:1001700002301420023086068316861001308312FF... 
:10017000023014200230860683168610013083126
:1001800086068316061064308312 4E20B30F9F2814
:100190009421291886142910 8610 2A18061 42A1C52
:1001A00061029089200930102308E008F01702002
:1001B0009E0011089F002A081E07A8002808033C7B
:1001C000031CEA2800308A0028088207832927298F
:1001D0005529EA280630A4001030A2005A308E00BB
:1001E00013014205 0308E00023014207D308E00FB
:1001F00013014205A308E000230142064308E00FA
:1002000023014200130B30 06400033033020318BD
:1002100 01B29AA30AE00B8217830AC007830B1008C
:10022000B8216430AE00B821A030AC00A030B100DD
:10023000B821B30F042926088C0027088D003F2021
:100240000C08A6000D08A7006400261855290630E2
:10025000A4001030A2008C308E00013014205030E9
:100260008E00023014207D308E00013014205F306B
:100270008E00023014200130B300640003303302DA
:10028000031854 29AA30AE00B8217830AC00A03051
:10029000B100B8216430AE00B821A030AC00783095
:1002A000B100B821B30F3D29C8280630A400103092
:1002B000A2008C308E00013014207 8308E00023085
:1002C00014206E308E00013014206 4308E00023015
:1002D00014200130B300 6400033033020318822974
:1002E000AA30AE00B821A030AC007830B100B821FF
:1002F0006430AE00B8217830AC00A030B100B82135
:100300 00B30F6B29C828AA30AE00B821A030AC00CA
:10031000A030B100B8216430AE00B8217830AC0014
:100320007830B100B821C828AB01B001A901AA01F9
:100330000130B300640033082F02031CA9296400B4
:10034000619A329AB0A64008619A729B00AB30FBE
:100350009A2964002D082B02031CB0290130A90042
:10036000640032083002031CB7290130AA000800DB
:100370000130B40064000B3034020318D8292E0871
:100380008C008D0106308400803001202C088C0008
:100390008D0106308400403001203108 8C008D0131
:1003A00063084002030012 00D304E20B40FBA29D1
:0603B00008006300D929DA
:02400E00F53F7C
:00000001FF
```

After the PIC 16F84 has been programmed and inserted into the I.C. socket on the robot's controller board, place the robot on the floor in a room with an open area so that the robot can walk. Place a few obstacles around the room to test the robot's obstacle avoidance capabilities.

When the robot is powered up, it will start to walk in a forward direction until it comes close to an object. Figure 10.9 shows Insectronic approaching a rock.

FIGURE 10.9 Insectronic approaching an obstacle.

If the sensed object is to the left of the robot, it will turn itself to the right and then check to see if it has cleared the obstacle. If not, it will turn a little more to the right and then continue to walk forward. The robot's navigational behavior is illustrated in Figure 10.10, where the possible walking paths of the robot are shown with the dotted arrows.

In the situations where an obstacle is met straight on, the robot backs up and then makes a random decision to turn to the left or right. If the wrong direction was chosen, it might take the robot an extra two moves to free itself. With this obstacle-avoidance scheme, it is interesting to see what happens when the robot encounters the corner of a room. It has no trouble navigating its way out within two or three direction changes.

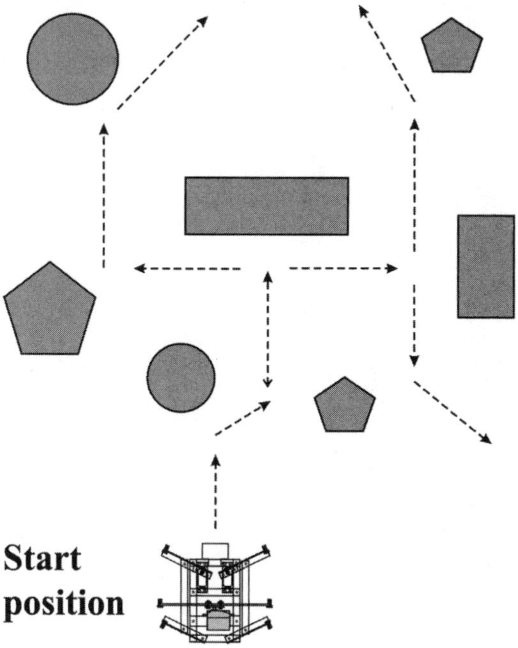

FIGURE 10.10 **Possible walking paths when obstacles are encountered.**

INSECT SEARCH BEHAVIOR WITH OBSTACLE AVOIDANCE

In Program 10.10, the robot would immediately continue to walk forward after sensing and avoiding an object. The next control program, listed in Program 10.11, is called `search-avoid.bas`, and will do a second sensor reading after an object has been detected. It will adjust the robot's position accordingly. If the sensed object is to the right of the robot, it will walk a few steps backwards, turn itself to the left, and then check to see if it has cleared the obstacle. If not, it will turn to the right further than its original position, and then continue to walk forward. The corresponding `search-avoid.hex` file is listed in Program 10.12. You will notice that the robot can better navigate using this scheme and does not get stuck as often. Another feature of this version is that the program can be put into the infrared calibration mode by connecting pins 2 and 3 of the mode select jumper on the controller board as shown in Figure 10.11. If you find that you need to readjust the infrared sensors, connect a jumper with the power turned off. When the robot is powered up with pins 2 and 3 connected, it will be running the code listed in Program 10.3 in a continuous loop. When you are finished making the adjustment, either remove the jumper or place it across pins 1 and 2.

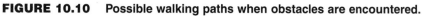

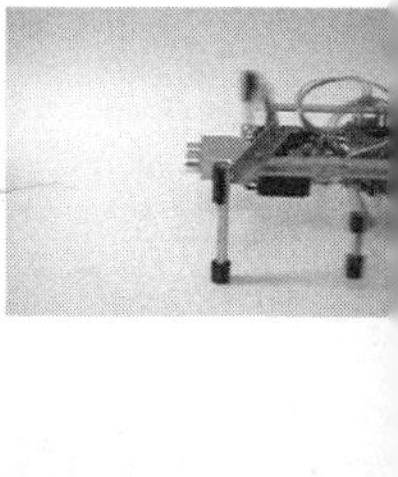

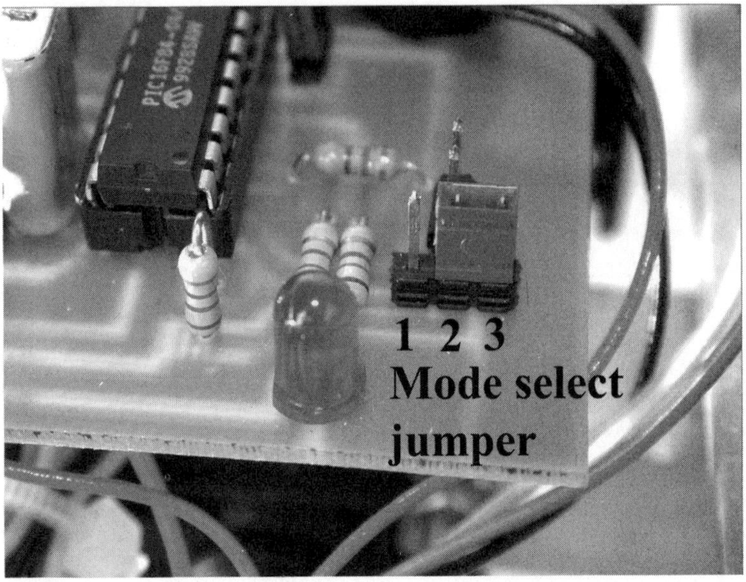

FIGURE 10.11 Mode select jumper set for calibration of the infrared sensors.

PROGRAM 10.11 Search-avoid.bas program listing.

```
'*****************************************************
' search-avoid.bas
' exploration with intelligent obstacle avoidance
' PicBasic Pro Compiler
'*****************************************************

trisa = %11111111                  ' set porta to inputs
trisb = %00001100                  ' set portb pins 2 & 3 to inputs

m_servo  var byte                  ' initialize variables
l_servo  var byte
r_servo  var byte
temp     var byte
temp2    var byte
timer    var byte
decision var word
ir_left  var byte
ir_right var byte
l_count  var byte
r_count  var byte
l_threshold var byte
r_threshold var byte
```

```
num_samples var byte
behavior var byte

num_samples = 30        ' number of samples taken per i.r. module
l_threshold = 5         ' noise floor threshold for left detector
r_threshold = 5         ' noise floor threshold for right detector

for temp = 1 to 10                          ' delay for approximately 1
                                            ' second by
    sound portb.4,[90 + (temp*3),2]         ' flashing LED's and making
                                            ' sounds so that
    toggle portb.1                          ' the infrared detecor board
                                            ' has time to
    toggle portb.0                          ' stabalize on power up.
    pause 100
next temp

start:
      if portA.2 = 1 then ir_cal

      gosub infrared    ' get sensor values from infrared subroutine

      portb.1 = ir_left
      portb.0 = ir_right

      behavior = ir_left * 2 + ir_right

      if behavior = 0 then
         gosub walk_forward
      endif

      if behavior = 1 then
         gosub walk_reverse
         gosub turn_left
         gosub infrared
               if ir_right = 1 then
                  for temp2 = 1 to 2
                      gosub turn_right
                  next temp2
               endif
      endif

      if behavior = 2 then
         gosub walk_reverse
         gosub turn_right
         gosub infrared
               if ir_left = 1 then
```

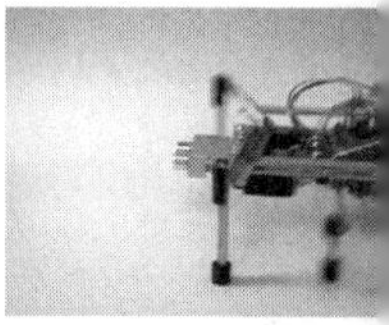

```
                    for temp2 = 1 to 2
                        gosub turn_left
                    next temp2
                endif
        endif

        if behavior = 3 then
            gosub walk_reverse
            gosub turn_left
            gosub infrared
                if ir_left = 1 then
                    for temp2 = 1 to 2
                        gosub turn_right
                    next temp2
                endif
        endif

goto start

walk_reverse:
                sound portb.4,[90,1,80,2,125,1,90,2,100,2]
                for temp = 1 to 2
                     m_servo = 170
                     gosub servo
                     l_servo = 120
                     r_servo = 120
                     gosub servo
                     m_servo = 100
                     gosub servo
                     l_servo = 160
                     r_servo = 160
                     gosub servo
                next temp
return

turn_left:
                sound portb.4,[140,1,80,2,125,1,95,2]
                for temp = 1 to 2
                     m_servo = 170
                     gosub servo
                     l_servo = 120
                     r_servo = 160
                     gosub servo
                     m_servo = 100
                     gosub servo
                     l_servo = 160
                     r_servo = 120
```

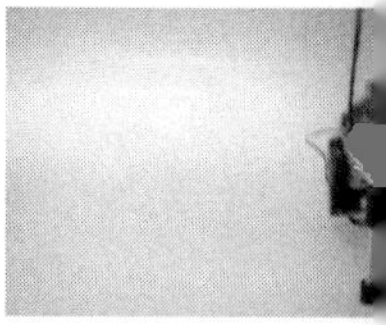

```
                gosub servo
            next temp
return

turn_right:
            sound portb.4,[140,1,120,2,110,1,100,2]
            for temp = 1 to 2
                m_servo = 170
                gosub servo
                l_servo = 160
                r_servo = 120
                gosub servo
                m_servo = 100
                gosub servo
                l_servo = 120
                r_servo = 160
                gosub servo
            next temp
return

walk_forward:
                m_servo = 170
                gosub servo
                l_servo = 160
                r_servo = 160
                gosub servo
                m_servo = 100
                gosub servo
                l_servo = 120
                r_servo = 120
                gosub servo
return

infrared:

l_count = 0
r_count = 0
ir_left = 0
ir_right = 0

        for temp = 1 to num_samples   ' take 30 samples per detector

            if portb.2 = 0 then          ' if left module senses an
                                         ' object
               l_count = l_count + 1     ' increase the left count
            endif
```

```
          if portb.3 = 0 then           ' if right module senses
                                        ' an object
             r_count = r_count + 1      ' increase the right count
          endif

      next

      if l_count >= l_threshold then     ' if left count is
                                         ' greater than the
                                         ' threshold
         ir_left = 1                     ' then an object was
                                         ' sensed with left
                                         ' detector
      endif

      if r_count >= r_threshold then     ' if right count is
                                         ' greater than the
                                         ' threshold
         ir_right = 1                    ' then an object was
                                         ' sensed with right
                                         ' detector
      endif

return                                  ' return to main program

servo:                                  ' subroutine to set servos

     for timer = 1 to 10
     pulsout portb.7,m_servo
     pulsout portb.6,l_servo
     pulsout portb.5,r_servo
     pause 13
     next timer

return

ir_cal:                      ' subroutine to calibrate I.R. sensors

    low portb.0

    If portb.3 = 0 then      ' Check right i.r. module output for
                             ' logic 0
        high portb.0         ' turn on right light emitting diode
    else
        low portb.0
    endif
```

```
        low portb.1

        If portb.2 = 0 then         ' Check left i.r. module output for
                                    ' logic 0
            high portb.1            ' turn on left light emitting diode
        else
            low portb.1
        endif

goto start

end
```

PROGRAM 10.12 **Search-avoid.hex file listing.**

```
:100000009128A2003B200C080D0403198C2886209F
:10001000841322088006640000D280E288C0A031918
:100020008D0F0B2880068C288F002408840022095D
:100030003C2084138F0803198C28F03091000E089F
:1000400080389000F03091030319910003198F0359
:1000500003198C282B283F2003010C1822088E1F1F
:1000600022088E0803190301900F382880061F28E4
:100070003928000002228FF3A841780058C280D08B3
:100080000C0403198C0A80300C1A8D060C198D068D
:100090008C188D060D0D8C0D8D0D8C288F018E000A
:1000A000FF308E07031C8F07031C8C2803308D0044
:1000B000DF305C2050288D01E83E8C008D09FC303B
:1000C000031C65288C07031862288C0764008D0FB9
:1000D00062280C186B288C1C6F2800006F28080001
:1000E000103094008D018C01930C920C031C7E281F
:1000F0000E088C070F0803180F0F8D078D0C8C0C42
:10010000910C900C940B742810088C288C098D0984
:100110008C0A03198D0A08008313031383126400E9
:10012000008008316FF3085000C30860083121E30D5
:10013000AF000530AD000530B2000130B3006400FF
:100140000B3033020318C8280630A4001030A20078
:100150003308920093010330 8E008F0170209E00BF
:1001600011089F005A309E0703189F0A1E088E0030
:10017000023014200230860683168610013083126 6
:100180008606831606106430831 24E20B30F9F2814
:100190006400051911 2ACC2129188614291C8610FF
:1001A0002A1806142A1C0610290892009301 02300E
:1001B0008E008F0170209E0011089F002A081E07E4
:1001C000A80064002808003C031DE728BB21640048
:1001D0002808013C031DFE282D215F21CC2164004D
```

```
:1001E0002A08013C031DFE280130B40064000330DE
:1001F00034020318FE288D21B40FF6286400280865
:10020000023C031D15292D218D21CC2164002908D4
:10021000013C031D15290130B40064000330340291
:10022000031815295F21B40F0D2964002808033C29
:10023000031D2C292D215F21CC2164002908013CBC
:10024000031D2C290130B4006400033034020318 6C
:100250002C298D21B40F2429C8280630A400103081
:10026000A2005A308E000130142050308E0002302F
:1002700014207D308E00013014205A308E00023060
:1002800014 2064308E00023014200130B30064006A
:10029000033 0330203185E29AA30AE00F021783013
:1002A000AC007830B100F0216430AE00F021A03015
:1002B000AC00A030B100F021B30F47290800063090
:1002C000A4001030A2008C308E0001301420503079
:1002D0008E00023014207D308E00013014205F30FB
:1002E0008E00023014200130B3006400033033026A
:1002F00003188C29AA30AE00F0217830AC00A03071
:10030000B100F0216430AE00F021A030AC007830B4
:10031000B100F021B30F752908000630A400103099
:10032000A2008C308E000130142078308E00023014
:10033000142 06E308E000130142064308E000230A4
:1003400014200130B3006400033033020318BA29CB
:10035000AA30AE00F021A030AC007830B100F0211E
:100360006430AE00F0217830AC00A030B100F02154
:10037000B30FA3290800AA30AE00F021A030AC00D2
:10038000A030B100F0216430AE00F0217830AC0034
:100390007830B100F0210800AB01B001A901AA0139
:1003A0000130B300640033082F02031CE12964000C
:1003B0000619DB29AB0A64008619DF29B00AB30FDE
:1003C000D22964002D082B02031CE8290130A90062
:1003D000640032083002031CEF290130AA00080033
:1003E0000130B50064000B3035020318102A2E08C6
:1003F0008C008D0106308400803001202C088C0098
:100400008D0106308400403001203108 8C008D01C0
:100410000630840020300120 0D304E20B50FF22927
:1004200008000610831606106400831286191D2A20
:10043000061483160 6108312212A0610831606104E
:1004400083128 6108316 86106400831206192D2AE3
:1004500086148 316 86108312312A86108316 86101E
:080460008312C8286300322A50
:02400E00F53F7C
:00000001FF
```

Further Experiments

The `search-avoid` program in this chapter gives the robot the ability effectively to navigate its way through a rough-terrain environment and avoid any obstacles that it encounters along the way. The reader now has enough information to experiment with programming new methods for reading and interpreting the signals from the infrared sensors, different walking gaits, control programs, and behaviors. For the robot to be able actually to map out the surrounding area, we will add an ultrasonic range finding system in Chapter 11. This system will give us far more navigational accuracy so that obstructions can be sensed at a distance instead of avoided at the last second.

A simple adaptation to the robot's walking subroutine that will speed up the gait and give the robot the appearance of being angry or agitated is accomplished by setting the servo values all at the same time. Although a delay is needed when changing the middle servo position, this is not too much of a problem because when the servo subroutine is called, the middle servo is always positioned first. For example, with the programs that we have written so far, a typical walking routine looks like this:

```
walk_forward:
                    m_servo = 170
                    gosub servo
                    l_servo = 160
                    r_servo = 160
                    gosub servo
                    m_servo = 100
                    gosub servo
                    l_servo = 120
                    r_servo = 120
                    gosub servo
return
```

Try modifying each of the walking routines by removing the servo subroutine call for the middle legs so that the code looks like this:

```
walk_forward:
                    m_servo = 170
                    l_servo = 160
                    r_servo = 160
                    gosub servo
                    m_servo = 100
                    l_servo = 120
                    r_servo = 120
                    gosub servo
return
```

Another experiment is to add in the servo call for the left servo so that each movement is very precise:

```
walk_forward:
                    m_servo = 170
                    gosub servo
                    l_servo = 160
                    gosub servo
                    r_servo = 160
                    gosub servo
                    m_servo = 100
                    gosub servo
                    l_servo = 120
                    gosub servo
                    r_servo = 120
                    gosub servo
return
```

The last program is called `running-bug.bas` and is listed in Program 10.13. It uses the fast running gait mentioned earlier and also includes the infrared module calibration routine that is controlled with the mode select jumper. Compile the code, program the 16F84 PIC with the `running-bug.hex` file listed in Program 10.14, and watch how fast it runs. I found that using this method makes the robot appear to be distressed and aggressive.

PROGRAM 10.13 **running-bug.bas program listing.**

```
'**************************************************************
' running-bug.bas
' exploration and obastacle avoidance with running gait
' PicBasic Pro Compiler
'**************************************************************

trisa = %11111111                    ' set porta to inputs
trisb = %00001100                    ' set portb pins 2 & 3 to inputs

m_servo  var byte                    ' initialize variables
l_servo  var byte
r_servo  var byte
temp      var byte
timer     var byte
decision var word
ir_left  var byte
ir_right var byte
```

```
l_count   var byte
r_count   var byte
l_threshold var byte
r_threshold var byte
num_samples var byte
behavior var byte

num_samples = 30          ' number of samples taken per i.r. module
l_threshold = 5           ' noise floor threshold for left detector
r_threshold = 5           ' noise floor threshold for right detector

for temp = 1 to 10                      ' delay for approximately 1
                                        ' second by
    sound portb.4,[90 + (temp*3),2]     ' flashing LED's and making
                                        ' sounds so that
    toggle portb.1                      ' the infrared detecor board
                                        ' has time to
    toggle portb.0                      ' stabalize on power up.
    pause 100
next temp

start:
      if portA.2 = 1 then ir_cal

      gosub infrared    ' get sensor values from infrared subroutine

      portb.1 = ir_left
      portb.0 = ir_right

      behavior = ir_left * 2 + ir_right

      branch behavior,[walk_forward, turn_left, turn_right,
  walk_reverse]

walk_reverse:
             sound portb.4,[90,1,80,2,125,1,90,2,100,2]
             for temp = 1 to 2
                 m_servo = 170
                 l_servo = 120
                 r_servo = 120
                 gosub servo
                 m_servo = 100
                 l_servo = 160
                 r_servo = 160
                 gosub servo
             next temp
```

```
                random decision

                if decision.0 = 1 then turn_right
turn_left:
                sound portb.4,[140,1,80,2,125,1,95,2]
                for temp = 1 to 2
                    m_servo = 170
                    l_servo = 120
                    r_servo = 160
                    gosub servo
                    m_servo = 100
                    l_servo = 160
                    r_servo = 120
                    gosub servo
                next temp
                goto start
turn_right:
                sound portb.4,[140,1,120,2,110,1,100,2]
                for temp = 1 to 2
                    m_servo = 170
                    l_servo = 160
                    r_servo = 120
                    gosub servo
                    m_servo = 100
                    l_servo = 120
                    r_servo = 160
                    gosub servo
                next temp
                goto start
walk_forward:
                    m_servo = 170
                    l_servo = 160
                    r_servo = 160
                    gosub servo
                    m_servo = 100
                    l_servo = 120
                    r_servo = 120
                    gosub servo
goto start

infrared:

l_count = 0
r_count = 0
ir_left = 0
ir_right = 0
```

```
          for temp = 1 to num_samples        ' take 30 samples per
                                             ' detector

               if portb.2 = 0 then           ' if left module senses an
                                             ' object
                  l_count = l_count + 1      ' increase the left count
               endif

               if portb.3 = 0 then           ' if right module senses
                                             ' an object
                  r_count = r_count + 1      ' increase the right count
               endif

          next

          if l_count >= l_threshold then    ' if left count is greater
                                            ' than the threshold
             ir_left = 1                    ' then an object was
                                            ' sensed with left detector
          endif

          if r_count >= r_threshold then    ' if right count is
                                            ' greater than the
                                            ' threshold
             ir_right = 1                   ' then an object was
                                            ' sensed with right
                                            ' detector
          endif

return                                      ' return to main program

servo:                                      ' subroutine to set servos

        for timer = 1 to 10
        pulsout portb.7,m_servo
        pulsout portb.6,l_servo
        pulsout portb.5,r_servo
        pause 13
        next timer

return

ir_cal:                          ' subroutine to calibrate I.R. sensors

       low portb.0

       If portb.3 = 0 then      ' Check right i.r. module output for
```

```
                              ' logic 0
        high portb.0          ' turn on right light emitting diode
    else
        low portb.0
    endif

    low portb.1

    If portb.2 = 0 then       ' Check left i.r. module output for
                              ' logic 0
        high portb.1          ' turn on left light emitting diode
    else
        low portb.1
    endif

goto start

end
```

PROGRAM 10.14 **running-bug.hex file listing.**

```
:100000009128A2003B200C080D0403198C2886209F
:10001000841322088006640000D280E288C0A031918
:100020008D0F0B2880068C288F002408840022095D
:100030003C2084138F0803198C28F03091000E089F
:10004000803890000F030910303199100003198F0359
:100050003198C282B283F2003010C1822088E1F1F
:1000600022088E08031903019000F382880061F28E4
:10007000392800002228FF3A841780058C280D08B3
:100080000C0403198C0A80300C1A8D060C198D068D
:100090008C188D060D0D8C0D8D0D8C288F018E000A
:1000A000FF308E07031C8F07031C8C2803308D0044
:1000B000DF305C2050288D01E83E8C008D09FC303B
:1000C00031C65288C07031862288C0764008D0FB9
:1000D00062280C186B288C1C6F2800006F28080001
:1000E00010309400 8D018C01930C920C031C7E281F
:1000F0000E088C070F0803180F0F8D078D0C8C0C42
:10010000910C900C940B742810088C288C098D0984
:100110008C0A03198D0A080083130313831264 00E9
:100120000080083 16FF3085000C30860083121E30D5
:10013000AF000530AD000530B2000130B3006400FF
:100140000B3033020318C8280630A4001030A20078
:1001500033089200930103308E008F0170209E00BF
:1001600011089F005A309E0703189F0A1E088E0030
```

```
:10017000023014200230860683168610013083126 6
:10018000860683160610643083124E20B30F9F2814
:100190000640005 19D4298F2129188614291C86107A
:1001A0002A1806142A1C061029089200930102300E
:1001B0008E008F0170209E0011089F002A081E07E4
:1001C000A8002808033C031CED2800308A002808FA
:1001D0008207802928295429ED280630A4001030F0
:1001E000A2005A308E000130142050308E000230B0
:1001F00014207D308E00013014205A308E000230E1
:10020000142064308E00023014200130B3006400EA
:100210000330330203181C29AA30AE007830AC003A
:100220007830B100B3216430AE00A030AC00A03013
:10023000B100B321B30F072926088C0027088D00D1
:100240003F200C08A6000D08A700640026185429BA
:100250000630A4001030A2008C308E000130142033
:100260000 50308E00023014207D308E00013014207A
:100270005F308E00023014200130B3006400033080
:100280000330203185329AA30AE007830AC00A030F6
:10029000B100B3216430AE00A030AC007830B100C2
:1002A000B321B30F3E29C8280630A4001030A200A5
:1002B0008C308E000130142078308E0002301420F3
:1002C0006E308E000130142064308E000230142015
:1002D0000130B3006400033033020318 7F29AA30D1
:1002E000AE00A030AC007830B100B3216430AE0075
:1002F0007830AC00A030B100B321B30F6A29C82810
:10030000AA30AE00A030AC00A030B100B321643000
:10031000AE007830AC007830B100B321C828AB0112
:10032000B001A901AA010130B300640033082F0213
:10033000031CA429640006199E29AB0A64008619CF
:10034000A229B00AB30F952964002D082B02031CC3
:10035000AB290130A900640032083002031CB22925
:100360000130AA0008000130B40064000B303402F0
:100370000318D3292E088C008D01063084008030AC
:1003800001202C088C008D01063 0840040300120B3
:1003900031088C008D01063084002030012 00D30A2
:1003A0004E20B40FB5290800061083160610640000D
:1003B00083128619E02906148316061083 12E42995
:1003C0000610831606108312861083168610640 0AA
:1003D00083120619F0298614831686108312F429D5
:0E03E000861083168610831 2C8286300F52944
:02400E00F53F7C
:00000001FF
```

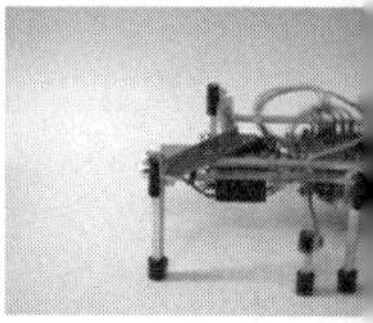

Summary

In this chapter, the robot was taught to walk and avoid obstacles using a number of different techniques. The limitations of using infrared detectors became apparent when the robot approached small objects. Another disadvantage of the infrared detectors is the small range in which they function, making room mapping impossible. In the next chapter, these limitations will be overcome by adding a sonar ranger. This device can detect small objects at a distance of 3 meters, making it ideal for room mapping. With the sonar ranger onboard, the robot will be able to make intelligent decisions about where to walk and store an internal representation of the outside world in its memory.

11

ULTRASONIC RANGE FINDING

In Chapter 10, the robot was able to avoid obstacles at close range using the infrared detection technique. This is effective for close reactive responses and has the advantage of a 180-degree detection radius to the front of the robot. But it does not allow the robot to look ahead to determine where it should walk *before* making a move based on the position of walls and obstacles. Another problem with the infrared detectors is that narrow objects like table legs are often missed unless encountered on the right angle. What is needed is a system that will be able to determine the distance of an object from the robot, and then make decisions based on that information. Another necessary requirement is for the robot to create a rudimentary map of the surrounding area before navigation through the environment begins. To meet these requirements, an ultrasonic range finding system will be implemented.

Devantech SRF04 Ultrasonic Range Finder

A low-cost solution is the Devantech SRF04 ultrasonic range finder pictured in Figure 11.1. This device offers precise ranging information from 3 cm to 3 meters, is easy to interface, and its minimal power requirements make it an ideal ranger for mobile robotics applications. It is available from Acroname Robotics Inc., and can be purchased from their Web site: http://www.acroname.com/.

FIGURE 11.1 Devantech SRF04 ultrasonic range finder.

The SRF04 range finder is a small printed circuit board that measures 1 3/4 × 3/4 inches with two ultrasonic transducers mounted on the front. The ranger requires a 5V power supply capable of handling roughly 50 mA of continuous output. One transducer is used to send an ultrasonic signal, and the other transducer receives the signal reflection from nearby objects. The SRF04 will output a 100-microsecond to 18-millisecond detection pulse that is proportional to range when a reflected signal is detected. Table 11.1 is a list of the parts that will be needed to add the sonar ranger. The SRF04 range finder specifications are listed in Table 11.2.

TABLE 11.1 List of Parts Required for the Addition of the SRF04 Ultrasonic Range Finder.

PART	QUANTITY	DESCRIPTION
SRF04 Ultrasonic ranger module	1	Sonar distance measuring device
5-post male header connector	1	2.5-mm spacing
4-strand ribbon cable	1	8 1/2 inches
2-connector female header	2	2.5-mm spacing
1/16-inch thick aluminum	1	2 inches × 4 inches
Hot glue	—	Hot glue gun is needed

TABLE 11.2 Table of Specifications for the SRF04

Voltage	5V
Current	30 mA Typical 50 mA
Frequency	40 kHz
Maximum range	3 meters
Minimum range	3 centimeters
Sensitivity	Can detect a 3-cm diameter broom handle at 2 meters
Input trigger	10 μS minimum TTL level pulse
Echo pulse	Positive TTL level signal, width proportional to range
Size	1 3/4 × 3/4 inches

THEORY OF OPERATION

The SRF04 works by sending a pulse of sound outside the range of human hearing. This pulse travels at the speed of sound (1.1 ft/millisecond) away from the ranger in a cone shape. If any objects are in the path of the pulse, the sound is reflected off the object and back to the ranger. The ranger is paused for a brief interval after the sound is transmitted, and then awaits the reflected sound in the form of an echo. The controller driving the ranger requests the device to create a 40-kHz sound pulse and then waits for the return echo. If the echo is received, the ranger reports this echo to the controller and the controller can then compute the distance to the object based on the elapsed time.

CONNECTIONS

The ranger requires four connections to operate. The first two are the power and ground lines. The ranger requires a 5V power supply capable of handling roughly 50 mA of continuous output. The other two lines are the signal connections. The first signal connection is the pulse trigger input line, and the second is the echo output line. These two pins will be connected to 2 I/O lines of the microcontroller. Figure 11.2 shows the connection pins on the back of the device. Note that the ground pin is on the far right and is marked with the letter "G" beside it.

BASIC TIMING

There are a couple of requirements to consider about the input trigger and the output pulse generated by the ranger. The input line should be held low (logic 0) and then

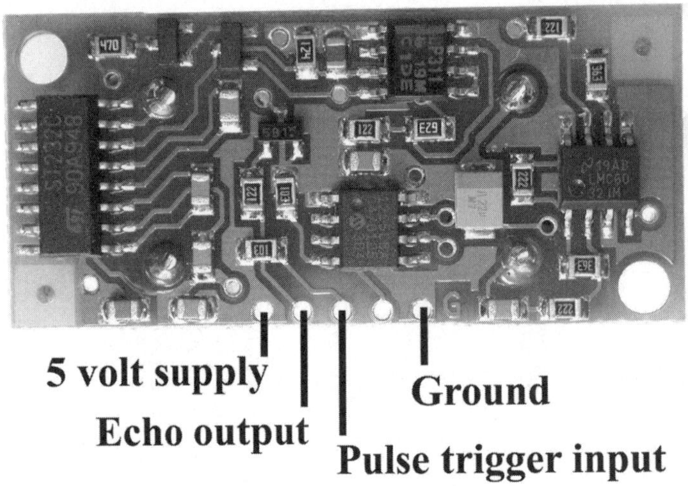

FIGURE 11.2 SRF04 pin connections.

brought high for a minimum of 10 μsec to initiate the sonic pulse. The pulse is generated on the falling edge of the input trigger. The ranger's *receive circuitry* is held in a short blanking interval of 100 μsec to avoid noise from the initial ping, and then it is enabled to listen to the echo. The echo line is logic low until the receive circuitry is enabled. Once the receive circuitry is enabled, the falling edge of the echo line signals either an echo detection or the timeout of 36 milliseconds if no object is detected. Figure 11.3 illustrates the timing sequence of the initial trigger input, the 40-kHz sonic burst that is generated, and the echo output pulse.

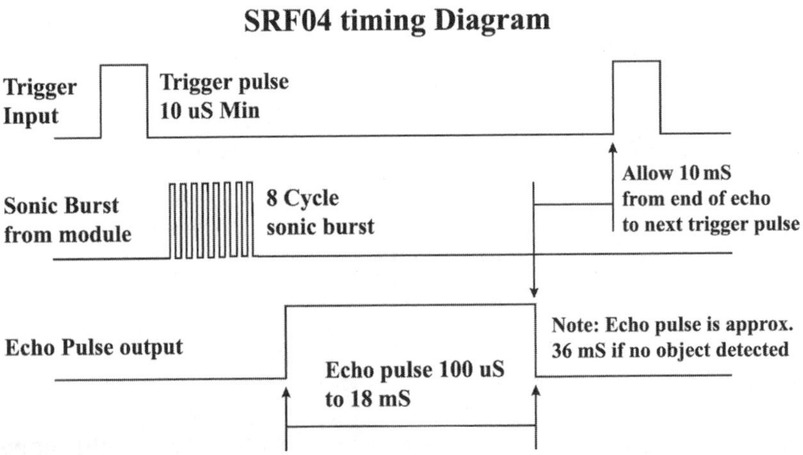

FIGURE 11.3 SRF04 timing diagram.

The microcontroller will begin timing on the falling edge of the trigger input pulse and end timing on the falling edge of the echo line. This duration determines the distance between the sonar module and the object from which the echo is bounced back. If no object is detected, a timeout will occur which is indicated by the echo output line going high for approximately 36 milliseconds.

Interfacing the SRF04 to the Robot's Controller Board

The ultrasonic ranger will replace the infrared detector board at the front of the robot. Unbolt the infrared sensor and disconnect the two jumper wires that connect the infrared board to the main controller circuit. Figure 11.4 shows how the SRF04 ultrasonic ranger is connected to the controller board.

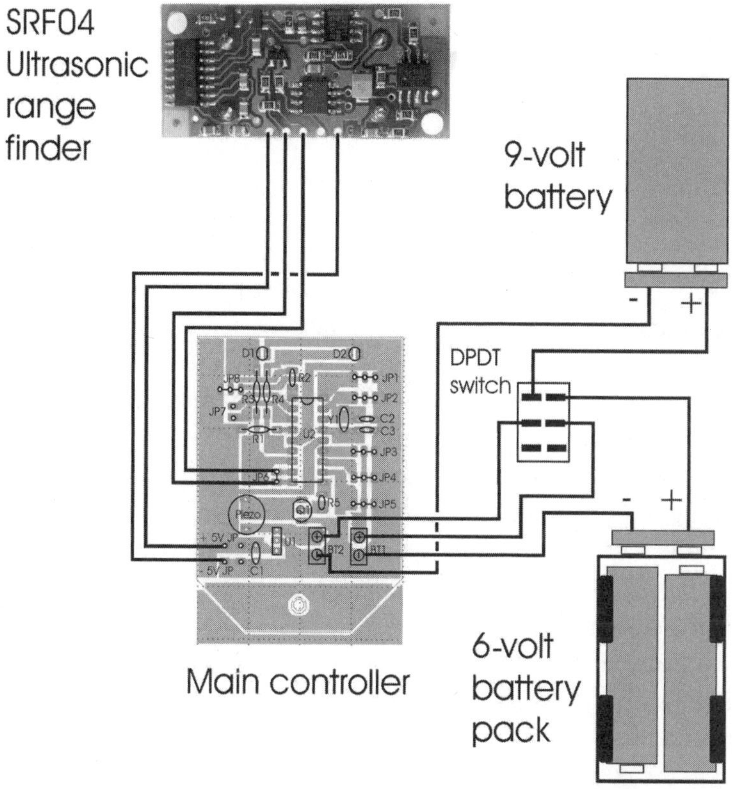

FIGURE 11.4 **SRF04 connections to the controller board.**

First, solder four male header pins to the ranger, as shown in Figure 11.5. This is probably the best way to connect the ranger to the controller since the robot could possibly move the wires around during walking. Wires soldered directly to printed circuit boards have a tendency to fray at the solder joints and become disconnected. The use of header pins eliminates this problem.

FIGURE 11.5 Header pins soldered to the SRF04 ultrasonic ranger.

Fabricate a jumper wire made up of 4-strand ribbon wire cut to a length of 8 1/2 inches. The end of the wire attached to the ranger uses a 5-connector female header. Solder the wires to the female header connector. Skip the pin that is not used and clip it off with wire cutters. On the other end of the wire, use a pair of 2-connector female headers and solder the 5V and ground to one connector and the trigger input and echo output to the other. Figure 11.6 illustrates what the connector wire should look like when it is finished.

To secure the ultrasonic ranger to the robot, we will fabricate a housing in which to mount the unit. Use Figure 11.7 as a guide to cut, drill, and bend the housing using 1/16-inch thick aluminum. Drill the mounting hole with a 5/32-inch drill bit. The aluminum can be bent on the edge of a table by hand or in a table vice. Figure 11.8 shows the finished housing, so that you can get an idea of how the aluminum should be bent. Next, place the ranger unit inside the housing at the front, and secure it in place by tightening the aluminum around the circuit board by hand. Apply a small amount of hot glue on the inside at the corners where the circuit board and aluminum housing meet. This will ensure that the circuit board does not move out of position.

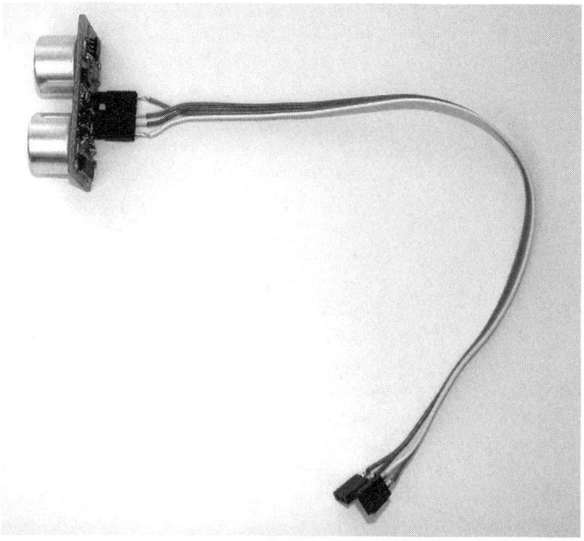

FIGURE 11.6 Completed SRF04 connector wire.

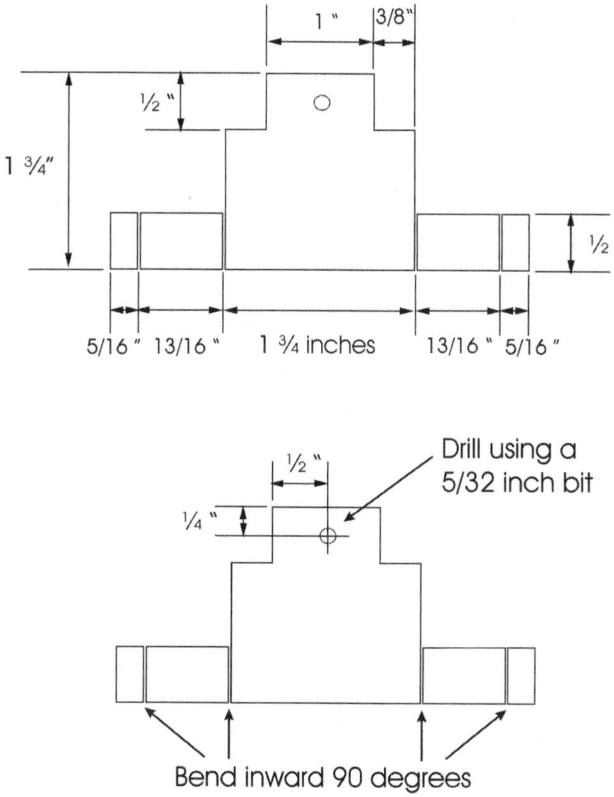

FIGURE 11.7 Cutting, drilling, and bending guide for the SRF04 housing.

FIGURE 11.8 Finished SRF04 housing.

Figure 11.9 shows the SRF04 ranger mounted in the housing with the jumper wire plugged into the header connector. The next step is to mount the sensor to the robot and connect the power, ground, trigger, and echo lines to the controller board as indicated in Figure 11.4. Make sure not to reverse the ranger's 5V and ground lines when plugging the connectors into the controller board. Note that the pulse trigger line is connected to the PORTB.2 pin, and the echo return is connected to the PORTB.3 pin. Mount the SRF04 housing to the robot's body where the infrared sensor module used to be. A single machine screw, washer, and bolt secure the ranger and 9V battery clip to the robot as shown in Figure 11.10. Make sure that a fresh a 9V battery is in place when the ranger housing is secured, since the housing needs to be unbolted to change the battery.

Using an LCD Serial Display

In order to write and test the software for the ultrasonic ranger module, the addition of a temporary LCD serial display will be discussed. If you choose not to add the LCD display, an alternative method for testing the module will also be covered. The ability to display the numeric information from the sensor module for coding and debug-

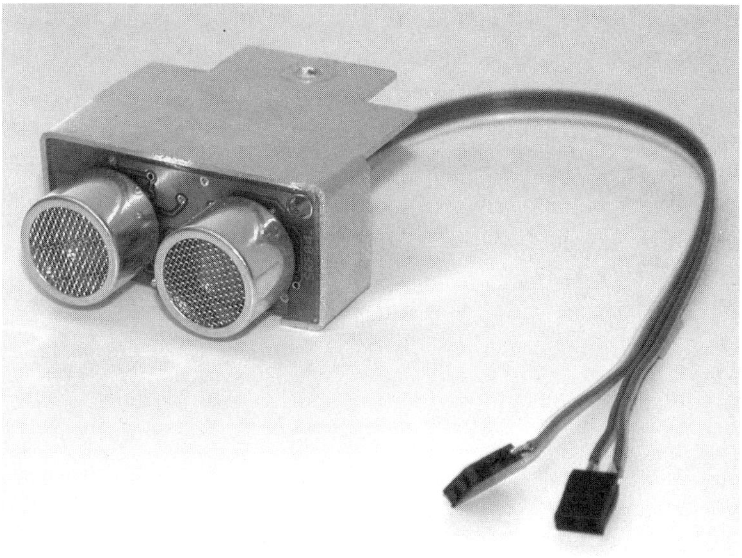

FIGURE 11.9 SRF04 ranger mounted in housing with connector wire attached.

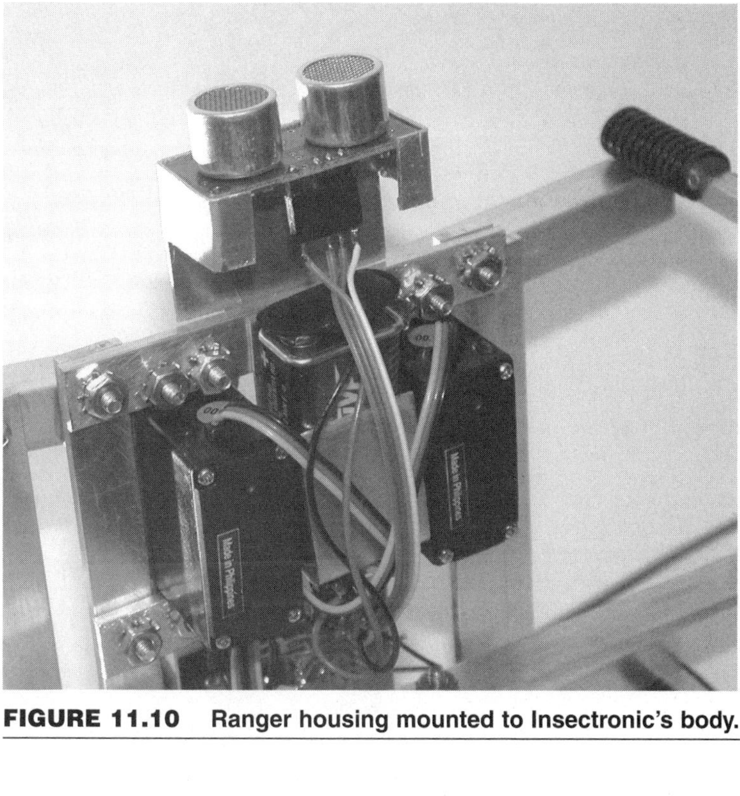

FIGURE 11.10 Ranger housing mounted to Insectronic's body.

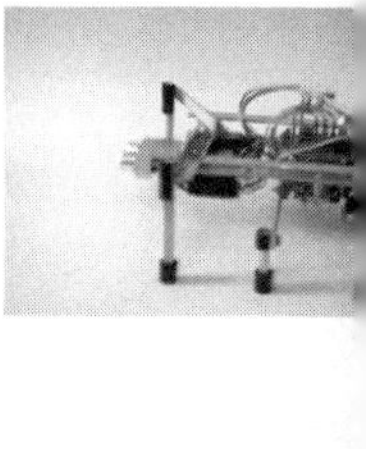

ging is valuable when developing software routines. The LCD display is not absolutely necessary for the addition of the SRF04 ranger module within the context of this book, but for serious experimentation, it is worth the small cost. The serial LCD module that will be discussed is called the *LCD Serial Backpack* and similar serial LCD displays can be purchased from the Scott Edwards Electronics Web site: http://www.seetron.com/. Figure 11.11 shows the LCD Serial Backpack attached to an LCD module with connectors attached.

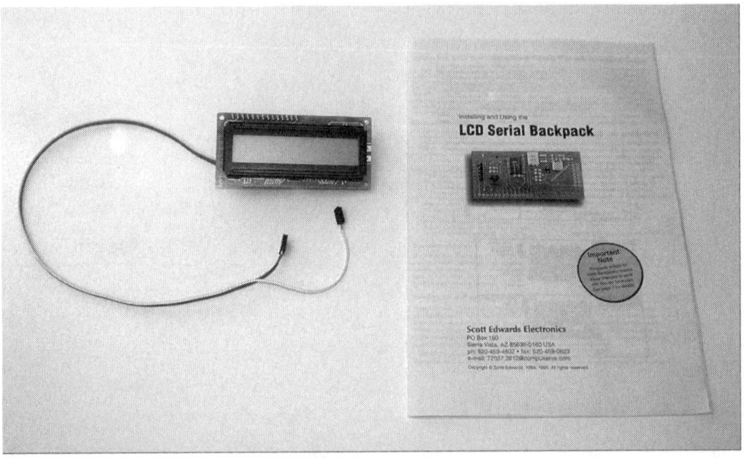

FIGURE 11.11 LCD Serial Backpack and LCD module.

Because LCD modules are so cheap and easy to find, the LCD Serial Backpack is a low-cost solution if you already have many LCD modules in your parts collection. The LCD Serial Backpack is a daughterboard that attaches to standard character LCD modules. It receives data serially and displays it on the LCD. The Backpack supports any alphanumeric LCD up to 80 screen characters (e.g., 4 lines $\times$ 20 characters). It accepts serial data at 2400 or 9600 baud (switch selectable). It is sold by itself, or preinstalled to high-quality 2×16 LCD modules. The Backpack has two modes: text and instruction. On power-up, the module defaults to text mode where any data sent to the Backpack is displayed on the screen. If the string “Hi Laurie” is sent, then “Hi Laurie” appears on the LCD. To distinguish text from instructions (e.g., clear screen, position cursor, etc.), the Backpack looks for an instruction prefix (ASCII 254). The byte following the prefix is treated as an instruction. After the instruction code, the Backpack returns to text mode. The specifications for the LCD Serial Backpack are listed in Table 11.3.

TABLE 11.3 Table of Specifications for the LCD Serial Backpack

Size	2.8 × 1.4 inch (71 × 36 mm)
Power Requirements	4.8 to 5.5Vdc @ < 1mA
User Connector	5-pin header; 0.025-inch posts on 0.10-inch centers
Connector Pinout	+5V GND SERIAL GND +5V
Serial Input	RS-232 or inverted TTL, 1200-9600, N81
Operating Temperature	0° to 50° C. For use only with commercial-temperature-range LCDs.
Initialization	Switches LCD power; performs soft init
Instruction Prefix	ASCII 254 (0FE hex)
LCDs Supported	Alphanumeric, to 80 characters (e.g., 4×20)

The detailed manual that is included with the Serial Backpack explains how to configure and use the device. Once the device is set up and configured (or if you buy an LCD display with the backpack already installed [BPI-216]), all you need to be concerned with is to supply power, ground, and a serial connection at the proper baud rate from the robot controller board.

CONNECTING THE LCD DISPLAY TO THE CONTROLLER BOARD

The parts required to add an LCD display are listed in Table 11.4.

TABLE 11.4 Parts List to Add an LCD Display.

PART	QUANTITY	DESCRIPTION
LCD Serial Backpack	1	LCD serial interface
2 × 16 LCD	1	Liquid crystal display
2-connector female header	1	2.5-mm spacing
1-connector female header	1	2.5-mm spacing
3-strand ribbon wire	1	14 inches
3-connector female header	1	2.5-mm spacing

Cut a piece of 3-strand ribbon wire to a length of 14 inches and solder the wires at one end to a 3-connector female header. On the other end, solder the two corresponding wires for 5V and ground to a 2-connector female header and the wire that connects to the serial input to a single connector female header. Use the connection diagram in Figure 11.12 to connect the LCD Serial Backpack to the controller.

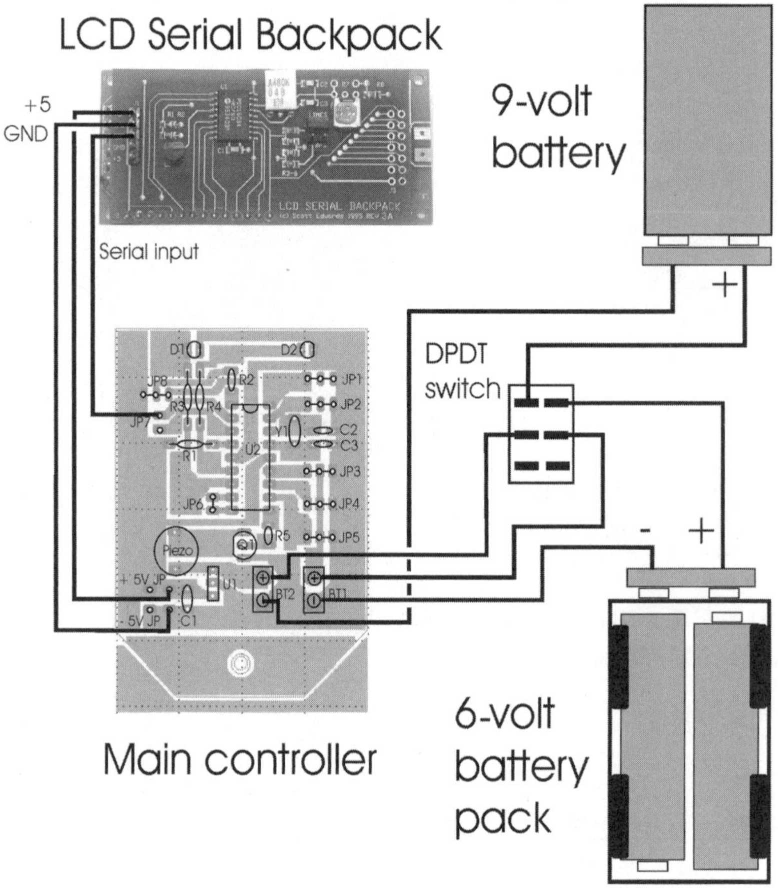

FIGURE 11.12 LCD Serial Backpack connection diagram.

Be very careful not to reverse the 5V and ground wires, or the LCD Backpack may become damaged. When the unit is attached to the robot, it should look like Figure 11.13. Having the LCD attached to the robot will only be a temporary arrangement for programming that will be removed when the robot is walking.

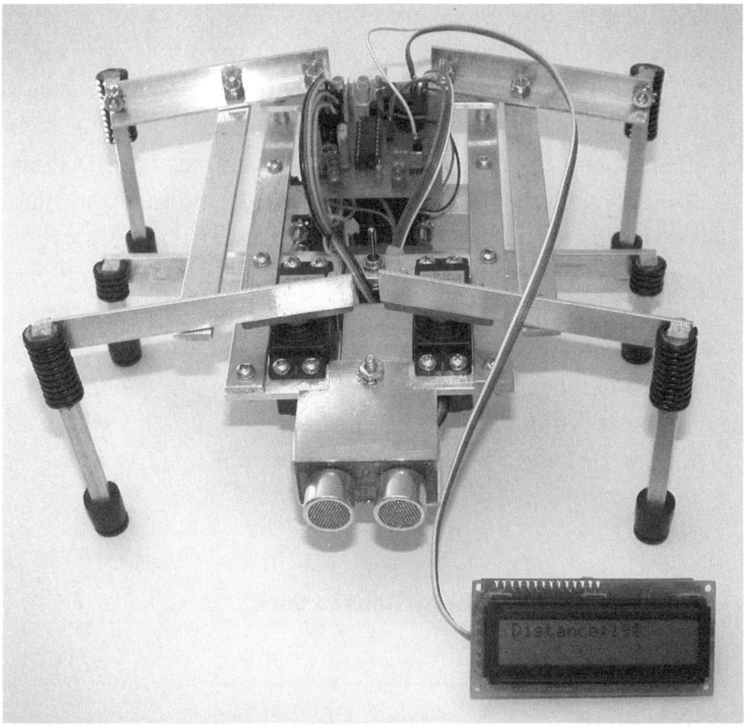

FIGURE 11.13 Serial LCD display attached to the controller board.

Testing the SRF04 Ultrasonic Ranger

As mentioned earlier, the ranger works by emitting a short burst of sound and then listening for the echo. Under the control of the PICmicro MCU 16F84, the SRF04 will emit an ultrasonic (40-kHz) sound pulse. The pulse travels through the air, hits an object, and then bounces back. Since we know that sound travels through air at approximately 1129 feet per second when the temperature is 21 degrees Celsius, we can accurately determine distance by measuring the amount of time between the transmission of the pulse and the return echo. When the temperature drops, the speed of sound through air slows down. If a temperature sensor were added, an algorithm to determine distance based on the speed of sound through air could take the surrounding temperature into account and adjust for differences.

The PicBasic Pro command called `pulsin` returns the round trip echo time in 10-μs units when using a 4-MHz oscillator. Since the pulse width has a 10-μs resolution per increment, that means that if `pulsin` returns a value of 1, then 10-μs have elapsed. The factors to convert the raw data to inches and centimeters given in the

SRF04 manual are 74 for inches (73.746 μs per 1 inch) and 29 for centimeters (29.033 μs per 1cm) based on the Basic Stamp's `pulsin` command returning values in 2-μs increments. In the SRF04 manual, the calculation to determine the distance is not divided in half to take into account the return time of the pulse because the sample program is for the Basic Stamp II, which returns `pulsin` values in 2-μs increments. Because the `pulsin` command with PicBasic Pro is returning values in increments of 10 μs, the conversion factor will need to be divided by 5 so that we get the correct value based on our 10-μs increment. Taking the `pulsin` increment timing difference into account gives us an approximate conversion factor of 15 for inches, and 6 for centimeters. Testing with the ranger indicated that the raw value returned by `pulsin` when an object was 12 inches away was 180. 180 divided by the inch conversion factor of 15 gives us the correct distance of 12 inches. Table 11.5 shows the I/O port pins of the PICmicro MCU 16F84 and the devices connected as reference when developing the software for the sonar ranger.

TABLE 11.5 List of Ranger and LCD I/O Connections to the PICmicro MCU 16F84

16F84 PORT	PIC CONFIGURATION	CONNECTION
PORTA.3	Output	LCD Serial Backpack input
PORTB.2	Output	SRF04 trigger input
PORTB.3	Input	SRF04 echo output
PORTB.0	Output	Right LED
PORTB.1	Output	Left LED
PORTB.4	Output	Piezo buzzer
PORTB.5	Output	Right servo
PORTB.6	Output	Left servo
PORTB.7	Output	Middle servo

Using the information about how the SRF04 operates, we can easily write a software routine to control the sending of the sonic pulse and then measure the pulse width of the return echo. The `sonar-ranger.bas` program uses the LCD to display the distance in inches and centimeters that an object is from the ranger. To send information to the LCD Serial Backpack with PicBasic Pro, the SEROUT command is used. The SEROUT command will send information to the specified output pin in standard asynchronous format using eight data bits, no parity, and one stop bit.

Create the `sonar-ranger.bas` program listed in Program 11.1 with a text editor, then save and compile with the PicBasic Pro compiler. Program the PICmicro

MCU 16F84 using the `sonar-ranger.hex` file along with the EPIC programmer. The `sonar-ranger.hex` file is listed in Program 11.2.

PROGRAM 11.1 **Sonar-ranger.bas program listing.**

```
'****************************************************************
' Sonar-ranger.bas
' Program to control the Devantech SRF04 ultrasonic module. Measure
' the distance between the the unit and an object. Convert the raw
' data to inches and centimeters and then send the results to an LCD
' display. The conversion factors listed in the SRF04 manual are for
' the PULSIN of a Basic Stamp II which measures in 2us increments.
' When running a PICmicro MCU at 4 MHz and using the PicBasic Pro
' compiler the PULSIN command measures in 10us increments and the
' PULSOUT command also generates pulses with 10us increments. The
' adjusted conversion factors for 10us increments are:
' inches = raw_data / 15
' centimeters = raw_data / 6
' PicBasic Pro Compiler by MicroEngineering Labs.
'****************************************************************

Include "modedefs.bas"              ' Include serial modes

trisa = %00000000                   ' set porta to outputs
trisb = %00001000                   ' set portb pin 3 to input

lcd         var     PORTA.3         ' define variables and constants
piezo       var     PORTB.4
trigger     var     PORTB.2
echo        var     PORTB.3
right_led   var     PORTB.0
left_led    var     PORTB.1
baud        con     N2400
dist_raw    var     word
dist_inch   var     word
dist_cm     var     word
conv_inch   con     15
conv_cm     con     6

low trigger                         ' set trigger pin to logic 0
low left_led                        ' turn off left LED
low right_led                       ' turn off right LED

Serout lcd,baud,[254,1]             ' clear lcd screen
```

```
pause 1000                        ' give time to initialize lcd

Serout lcd,baud,[254,128,"- Sonar Ranger -"]  ' display program
                                              ' title on the LCD

sound piezo,[100,10,50,5,70,10,50,2]          ' Make startup sound

pause 1000                                    ' wait for 1 second

Serout lcd,baud,[254,128,"inches:      "]   ' set up the LCD display

Serout lcd,baud,[254,192,"centimeters:  "]

main:

    gosub sr_sonar                            ' get sonar reading

    Serout lcd,baud,[254,135,#dist_inch,"  "] ' display the
                                              ' distance in inches

    Serout lcd,baud,[254,204,#dist_cm,"   "]  ' display the
                                              ' distance in
                                              ' centimeters

Goto main

end

sr_sonar:

     pulsout trigger,1                  ' send a 10us trigger pulse
                                        ' to the SRF04
     pulsin echo,1,dist_raw             ' start timing the pulse
                                        ' width on echo pin
     dist_inch = (dist_raw/conv_inch)   ' Convert raw data into
                                        ' inches
     dist_cm = (dist_raw/conv_cm)       ' Convert raw data into
                                        ' centimeters
    pause 1                             ' wait for 10us before
                                        ' returning to main

return
```

PROGRAM 11.2 Sonar-ranger.hex file listing.

```
:100000000F29A0008417800484138E010C1C8E001D
:1000100023200319OA2923200319OA2923200A2946
```

```
:10002000A000B3200C080D0403190A29042184132D
:10003000200880066400 1C281D288C0A03198D0FD7
:100040001A2880060A2920088E0601308C008D01AE
:10005000000820050E06031D08008C0A03198D0FE9
:1000600028280800910190009F1727308F0010303A
:10007000452003308F00E83045208F016430452053
:100080008F010A30452010085028 8E0011088D007D
:1000900010088C00E8200C08031D9F139F1B08000C
:1000A000303E92002208840009309300031 05D2046
:1000B000920C930B572803145D2884139F1D6C2802
:1000C000000820041F1D200680008417000820045B
:1000D000031C2006800077280008200403 1C20064B
:1000E0001F192006800084172009800577281F0D1E
:1000F00006398C0080208D008C0A80200000D428D6
:100100000308A000C0882070134753403341534 3A
:1001100000343C340C34D9348F0022088400200988
:10012000B42084138F0803190A29F03091000E08B7
:1001300080389000F030910303199100031 98F0368
:1001400003190A29A328B72003010C1820088E1FC1
:1001500020088E0803190301900FB0288006972805
:10016000B12800009A28FF3A841780050A290D0853
:100170000C0403198C0A80300C1A8D060C198D069C
:100180008C188D060D0D8C0D8D0D0A298F018E009A
:10019000FF308E07031C8F07031C0A2903308D00D4
:1001A000DF30D420C8288D01E83E8C008D09FC305A
:1001B000031CDD288C070318DA288C0764008D0FD8
:1001C000DA280C18E3288C1CE7280000E728080030
:1001D0009101900110309 2000D0D900D910D0E08BF
:1001E00090020F08031C0F0F91020318FE280E083F
:1001F00090070F0803180F0F910703108C0D8D0D3A
:10020000920BEC280C080A298C098D098C0A031919
:100210008D0A0800831303138312640008008316F9
:100220008501083086008312061183160611831299
:10023000061083160610831286108316861005306A
:100240008312A2000830A00004309F00FE3051202D
:1002500001305120033 08F00E830C7200530A20064
:100260000830A00004309F00FE305120803051202 3
:100270002D305120203051205330512 06F305120EB
:100280006E305120613051207230512020305120 89
:10029000523051206130512 06E305120673051205 2
:1002A000653051207230512020305120 2D305120A6
:1002B0000630A2001030A00064308E000A308C207E
:1002C00032308E0005308C2046308E000A308C2073
:1002D00032308E0002308C2003308F00E830C7208F
:1002E0000530A2000830A00004309F00FE305120ED
:1002F00080305120693051206E30512063305120C0
:10030000683051206530512073305120 3A305120EF
```

```
:10031000203051202030512020305120203051200D9
:10032000203051202030512020305120203051200C9
:100330002030512005530A2000830A00004309F007A
:10034000FE305120C0305120633051206530512065A3
:100350006E305120743051206930512066D305120651
:100360006530512074305120653051207230512059
:1003700073305120A3305120203051202030051200C
:10038000203051202030512F0210530A2000830CB
:10039000A00004309F00FE3051208730512027058F4
:1003A0009100260833202030512020305120203069
:1003B000051200530A2000830A00004309F00FE301C
:1003C0000005120CC3051202508910024083320203000C2
:1003D0000005120203051202030512020C4296300EE29C3
:1003E00001308C008D0106308400004301020013073
:1003F0008C0006308400083001200C08A8000D088D
:10040000A90028088C0029088D000F308E008F016C
:10041000E820A6000D08A70028088C0029088D00F8
:1004200006308E008F01E820A4000D08A50013000E1
:04043000C6200800DA
:02400E00F53F7C
:00000001FF
```

Once the PicMicro MCU 16F84 is programmed, insert the chip into the main controller board and turn on the power switch. The LCD will display '- Sonar Ranger-', play some notes on the piezo element, and then start measuring and displaying the distance of objects from the ranger in inches and centimeters as shown in Figure 11.14. Place a ruler in front of the ranger module and move your hand away from the unit in one-inch increments. If all goes well, the distance shown on the LCD display should correspond with the distance on the ruler.

DISTANCE REPRESENTED AS TONES

In the next experiment with the SRF04 sonar ranger, an audible tone will be produced based on the distance of an object from the device. The code listed in Program 11.3 is called range-tone.bas and does not use the LCD serial display. This program should be used to test the module if you do not have an LCD display. Compile the range-tone.bas code and program the PICmicro MCU 16F84 with the range-tone.hex file listed in Program 11.4. When the PIC is inserted into the socket and power is applied, move your hand slowly towards the ranger and notice that the tones produced by the PIC get lower the closer your hand gets to the device. When your hand is within 5 inches of the front of the ranger, both the left and right light-emitting diodes will turn on.

FIGURE 11.14 LCD displaying inches and centimeters.

PROGRAM 11.3 Range-tone.bas program listing.

```
'****************************************************************
' Range-tone.bas
' Program to control the Devantech SRF04 ultrasonic module. Measure
' the distance between the unit and an object. Convert the raw
' data to a frequency and output to the piezo element so that
' distance is represented by an audible frequency.
' Turn on the LEDs if an object is within 5 inches of the module.
' PicBasic Pro Compiler by MicroEngineering Labs.
'****************************************************************

trisa = %00000000                  ' set porta to outputs
trisb = %00001000                  ' set portb pin 3 to input

lcd          var     PORTA.3       ' define variables and constants
piezo        var     PORTB.4
trigger      var     PORTB.2
echo         var     PORTB.3
left_led     var     PORTB.0
right_led    var     PORTB.1
dist_raw     var     word
dist_inch    var     word
dist_cm      var     word
conv_inch    con     15
conv_freq    con     6
freq         var     word

low trigger                        ' set trigger pin to logic 0
low left_led                       ' turn off left LED
low right_led                      ' turn off right LED

main:
```

```
    gosub sr_sonar              ' get sonar reading
    if freq > 47 then main
    sound piezo,[80 + freq,10]  ' output frequency determined by
                                ' distance
    if dist_inch <= 5 then      ' if an object is 5 inches or less
        high left_led           ' away from the ranger then turn on
        high right_led          ' the left and right LEDs
    else
        low left_led            ' turn off the left and right LEDs
        low right_led
    endif

Goto main

end

sr_sonar:

        pulsout trigger,1                   ' send a 10us trigger
                                            ' pulse to the SRF04
        pulsin echo,1,dist_raw              ' start timing the
                                            ' pulse width on echo
                                            ' pin
        dist_inch = (dist_raw/conv_inch)    ' Convert raw data
                                            ' into inches
        freq = (dist_raw/conv_freq)         ' Convert raw data
                                            ' into a frequency
        pause 1                             ' wait for 10us before
                                            ' returning to main

return
```

PROGRAM 11.4 Range-tone.hex file listing.

```
:10000000C828A2008417800484138E010C1C8E0063
:1000100023200319C32823200319C3282320C3281E
:10002000A20059200C080D040319C328BD20841315
:100030002208800664001C281D288C0A03198D0FD5
:100040001A288006C32822088E0601308C008D01F4
:1000500000822050E06031D08008C0A03198D0FE7
:100060002828080008F002408840022095A208413BD
:100070008F080319C328F03091000E0880389000D3
:10008000F0309103031991000319 8F030319C3285A
:1000900049285D2003010C1822088E1F22088E08B3
```

```
:1000A00003190301900F562880063D2857280000A9
:1000B0004028FF3A84178005C3280D080C04031953
:1000C0008C0A80300C1A8D060C198D068C188D0642
:1000D0000D0D8C0D8D0DC3288F018E00FF308E0706
:1000E00031C8F07031CC32803308D00DF307A20E8
:1000F0006E288D01E83E8C008D09FC30031C83289E
:100100008C07031880288C0764008D0F80280C183A
:1001100089288C1C8D2800008D2808008E00013055
:10012000912894000F080D02031D98280E080C0258
:1001300004300318013003190230140503 1DFF3089
:10014000C328910190011030920000D0D900D910D7A
:100150000E0890020F08031C0F0F91020318B72816
:100160000E0890070F0803180F0F910703108C0D4E
:100170008D0D920BA5280C08C3288C098D098C0ABB
:1001800003198D0A08008313031383126400080007
:10019000831685010830860083120611831606112 6
:1001A0008312061083160610831286108316861 09B
:1001B00083120F212C088C002D088D008F012F3009
:1001C0008E20031DD9280630A4001030A200503024
:1001D0002C079E002D080318013E9F001E088E006C
:1001E0000A30322028088C0029088D008F01053044
:1001F0008E20031D0429061483160610831286140C
:10020000831686108312 0C29061083160610831 29B
:1002100086108316861083 12D92863000D290130B9
:100220008C008D010630840004301020013 08C00D9
:100230000630840008300120 0C08AA000D08AB002D
:100240002A088C002B088D000F308E008F01A12012
:10025000A8000D08A9002A088C002B088D00063084
:100260008E008F01A120AC000D08AD0001306C2084
:020270000080084
:02400E00F53F7C
:00000001FF
```

Now that the microcontroller is capable of accurately measuring distances with the SRF04 sonar ranger, the distance information can be incorporated into control programs. The next program is called `radar-walk.bas` and is listed in Program 11.5. Compile the code and program the Picmicro 16F84. This experiment will demonstrate how the robot can use the information received from the sonar ranger to keep a predetermined distance away from objects. Set the robot on the floor with nothing in front of it and turn on the power. Hold your hand in front of the robot about 30 cm away from the sonar ranger. Let the robot walk towards your hand and notice that it stops walking at approximately 15 cm. Move your hand towards the stationary robot and it will back away from you. The robot is trying to keep a constant distance between itself and any objects.

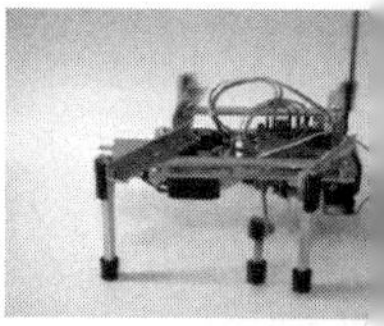

PROGRAM 11.5 Radar-walk.bas program listing.

```
'****************************************************************
' radar-walk.bas
' robot maintains a constant distance from an object.
' If the object is within 15 to 20 centimeters then the
' robot will stand still. If an object is closer than
' 15 cm then the robot will back away. If an object is
' beyond 20 cm then the robot will move towards it.
' PicBasic Pro Compiler
'****************************************************************

trisa = %00000000                   ' set porta to outputs
trisb = %00001000                   ' set portb pin 3 to input

m_servo      var   byte             ' define variables and constants
l_servo      var   byte
r_servo      var   byte
timer        var   byte
right_led    var   PORTB.0
left_led     var   PORTB.1
trigger      var   PORTB.2
echo         var   PORTB.3
piezo        var   PORTB.4
right_servo  var   PORTB.5
left_servo   var   PORTB.6
mid_servo    var   PORTB.7
dist_raw     var   word
dist_inch    var   word
dist_cm      var   word
conv_inch    con   15
conv_cm      con   6

low trigger                         ' set trigger pin to logic 0
low left_led                        ' turn off left LED
low right_led                       ' turn off right LED

main:

     gosub sr_sonar                               ' get sonar reading

     if dist_cm >= 15 and dist_cm <= 20 then      ' if an object is
                                                  ' within 15 to
          low left_led                            ' 20 cm then stand
                                                  ' still

          low right_led
          goto main
```

```
      endif

      if dist_cm > 20 then          ' if an object is less than 10 cm
           high left_led            ' turn on left LED
           low right_led            ' turn off right LED
           gosub walk_forward       ' walk towards object
      endif

       if dist_cm < 15 then
           low left_led             ' turn off left LED
           high right_led           ' turn on right LED
           gosub Walk_reverse       ' walk away from object
      endif

Goto main

end

walk_forward:
     sound portb.4,[90,1,80,2,125,1,90,2,100,2]
     m_servo = 170
     gosub servo
     l_servo = 160
     r_servo = 160
     gosub servo
     m_servo = 100
     gosub servo
     l_servo = 120
     r_servo = 120
     gosub servo
return

walk_reverse:
     sound portb.4,[90,1,80,2,125,1,90,2,100,2]
     m_servo = 170
     gosub servo
     l_servo = 120
     r_servo = 120
     gosub servo
     m_servo = 100
     gosub servo
     l_servo = 160
     r_servo = 160
     gosub servo
return

sr_sonar:
```

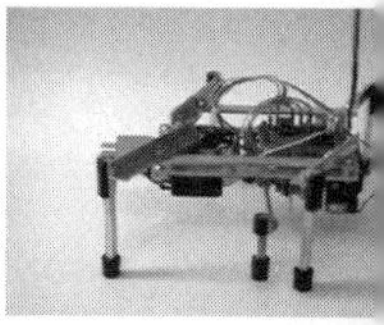

```
        pulsout trigger,1              ' send a 10us trigger
                                       ' pulse to the SRF04
        pulsin echo,1,dist_raw         ' start timing the
                                       ' pulse width on echo
                                       ' pin
        dist_inch = (dist_raw/conv_inch)  ' Convert raw data
                                       ' into inches
        dist_cm = (dist_raw/conv_cm)   ' Convert raw data
                                       ' into centimeters
        pause 1                        ' wait for 10us before
                                       ' returning to main

return

servo:                                 ' subroutine to set servos

        for timer = 1 to 10
        pulsout mid_servo,m_servo
        pulsout left_servo,l_servo
        pulsout right_servo,r_servo
        pause 13
        next timer

return
```

PROGRAM 11.6   **Radar-walk.hex file listing.**

```
:10000000D228A4008417800484138E010C1C8E0057
:1000100023200319CD2823200319CD282320CD2800
:10002000A40059200C080D040319CD28C7208413FF
:100030002408800664001C281D288C0A03198D0FD3
:100040001A288006CD2824088E0601308C008D01E8
:10005000000824050E06031D08008C0A03198D0FE5
:1000600028280800 8F002608840024095A208413B9
:100070008F080319CD28F03091000E0880389000C9
:10008000F03091030319910003198F030319CD2850
:1000900049285D2003010C1824088E1F24088E08AF
:1000A00003190301900F562880063D2857280000A9
:1000B0004028FF3A84178005CD280D080C04031949
:1000C0008C0A80300C1A8D060C198D068C188D0642
:1000D0000D0D8C0D8D0DCD288F018E00FF308E07FC
:1000E000031C8F07031CCD2803308D00DF307A20DE
:1000F0006E288D01E83E8C008D09FC30031C83289E
:100100008C07031880288C0764008D0F80280C183A
```

```
:1001100089288C1C8D2800008D2808008E00033053
:1001200094288E000630942894000F080D02031DB9
:100130009B280E080C02043003180130031902300A
:100140001405031DFF30CD280038031DFF300405C2
:10015000031DFF30CD2891019001103092000D0D4C
:10016000900D910D0E0890020F08031C0F0F9102C5
:100170000318C1280E0890070F0803180F0F9107E6
:1001800003108C0D8D0D920BAF280C08CD288C0917
:100190008D098C0A03198D0A080083130313831237
:1001A000640008008316850108308600831206115A
:1001B000831606118312061083160610831286100A
:1001C000831686108312812128088C0029088D004F
:1001D0008F010F308E209E0028088C0029088D008A
:1001E0008F0114309120A0001E0884002008A42054
:1001F000A000A100640020082104031908290610AA
:100200008316061083128610831686108312E32845
:1002100028088C0029088D008F0114309120031DBF
:100220001A29061483160610831286108316861068
:1002300083122F2128088C0029088D008F010F3090
:100240008E20031D2C290610831606108312861497
:100250008316861083125821E32863002D29063067
:10026000A6001030A4005A308E00013032205030E9
:100270008E00023032207D308E00013032205A3024
:100280008E000230322064308E0002303220AA30DC
:10029000AF00AB21A030AE00A030B000AB21643085
:1002A000AF00AB217830AE007830B000AB21080051
:1002B0000630A6001030A4005A308E0001303220E3
:1002C00050308E00023032207D308E0001303220DE
:1002D0005A308E000230322064308E0002303220DC
:1002E000AA30AF00AB217830AE007830B000AB213F
:1002F0006430AF00AB21A030AE00A030B000AB2125
:10030000080001308C008D0106308400043010207C
:1003100001308C0006308400083001200C08AC004D
:100320000D08AD002C088C002D088D000F308E00BC
:100330008F01AB20AA000D08AB002C088C002D0803
:100340008D0006308E008F01AB20A8000D08A9009B
:1003500001306C2008000130B10064000B30310224
:100360000318CB292F088C008D01063084008030C3
:1003700010202E088C008D010630840040301020A3
:1003800030088C008D0106308400203010200D30A4
:080390006C20B10FAD2908003B
:02400E00F53F7C
:00000001FF
```

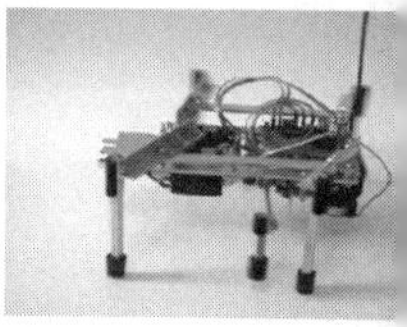

Room Mapping Using Sonar Range Finding

In Chapter 10, the robot avoided obstacles that it encountered by adjusting where it would walk based on the immediate sensor feedback. If an object were detected to the right side of the robot, then the robot would turn to the left and continue walking forward. The robot had no internal representation of the outside world around it—it just blindly wandered close to objects until the infrared sensors signaled that a response was needed. This is an effective way to keep the robot from bumping into objects, but it would be better if the robot could look ahead and not go near objects before walking. Now that the capabilities of an ultrasonic range finder have been added, the robot will be able to map out the area around itself and store the results internally. Based on this information, the robot can then make an intelligent decision about where to walk.

This is accomplished by having the robot take a series of distance measurements in a 180-degree arc to the front and sides of where it is standing before moving. When the robot is facing forward, it will rotate itself 90 degrees to the left and then take the first of eleven distance readings.

The SRF04 ultrasonic ranger has a detection cone of approximately 30 degrees, but accurate distance measurements are only obtained within 20 degrees or so. Because the robot is scanning in 18-degree slices for a total of 180 degrees around the front of the robot, there is some overlap in the detection cone for each slice. This ensures that there is never a "dead zone" where objects could be overlooked during the distance measuring routine.

The distance readings are stored in a one-dimensional array called position[12], made up of twelve elements. The first reading is stored in position[0]. The robot is then rotated to the right by 18 degrees and the second reading is stored in position[1]. This process continues until the robot has rotated through 180 degrees and has taken eleven distance measurements filling up the array elements position[0] through position[10]. Figure 11.15 illustrates the eleven positions that the robot rotates through taking and storing distance measurements in the array named position[12].

The last array element, position[11], is loaded with the value of zero and is needed when running a sort algorithm that determines which position has the largest distance measurement.

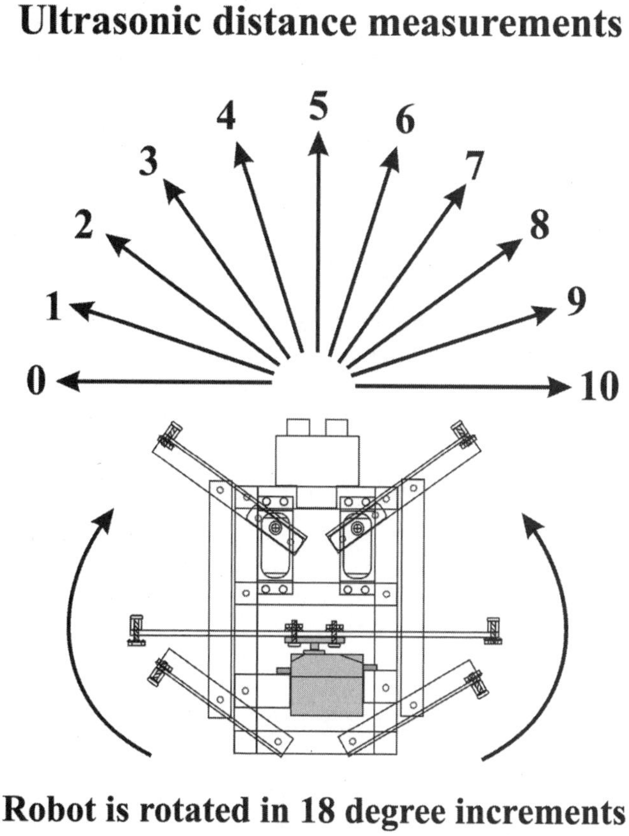

FIGURE 11.15 The robot takes eleven distance measurements while rotating on the spot.

The section of code to take the distance measurements and store each reading into the corresponding array element is as follows:

```
for turn = 1 to 5          ' rotate the robot 90 degrees to the left
  gosub turn_left
next turn

position[11] = 0           ' set last element of the array to zero

for i = 0 to 10            ' run through loop i 11 times
  gosub sr_sonar           ' get sonar reading dist_inch
  position[i] = dist_inch  ' store reading in array position[i]
  gosub turn_right         ' turn 18 degrees to the right
next i                     ' i = i + 1
```

Once the robot has rotated 90 degrees to the left, a for/next loop is initiated with the variable i. The first reading is stored in the array position[i], using i as the index. The first time through the loop i = 0. The robot is then rotated 18 degrees to the right. The variable i is incremented by the for/next loop and this continues until the upper limit of the for/next loop is reached. In this case, the body of the for/next loop will be executed eleven times.

When all the distance measurements have been taken, a sort algorithm will be applied to the values in the array to determine the area to the front and sides of the robot that has the most open space. Table 11.6 shows the position array filled with example distance measurements. It explains the function of the sorting algorithm that determines the location of the most space in front of the robot.

TABLE 11.6 Array "Position[12]" Filled with Example Sonar Measurements

Index	0	1	2	3	4	5	6	7	8	9	10	11
Distance measured	9	7	14	24	44	45	29	18	16	20	23	0

The sorting algorithm is:

```
best_pos = 11

for i = 0 to 10
if position[i] >= position[best_pos] then
best_pos = i
endif
next i

     most_space = 11 - best_pos
```

The variable called `best_pos` is used to keep track of the array index of the element containing the distance measurement with the highest value. The last array element, position[11], is loaded with the value of zero. A for/next loop is initiated with the variable `i`, starting with a value of 0. Variable i is used to index the array position[i]. The value stored in the array element position[i] is compared to the value stored in position `best_pos`. If the distance measurement stored in position[i] is larger than or equal to the value stored in position `best_pos`, then `best_pos` is updated with the current array index value of i. The variable `best_pos` starts off with the value of 11 so that on the first time through the loop when the comparison is done between the first array element and the last, the first element will always be chosen. If you work through the code segment with the values shown in Table 11.6, the value for the best position will be 5.

Once the best position has been determined, the robot will need to be rotated back to the position it was in when the measurement with the greatest distance was taken. To accomplish this, the value contained in variable `best_pos` is subtracted from 11 and stored in the variable `most_space`. If it is determined that the largest distance measurement is stored in position[10], the robot will still need to be rotated left 1 time to rotate into the correct position. The reason for this is that when the robot was taking the distance measurements with the for/next loop, the rotate-right subroutine was called after each measurement was taken. So, after the last measurement for position[10] was taken, the robot was rotated by 18 degrees to the right. The robot will then need to execute the `turn_left` subroutine the number of times determined by `most_space` to get itself into the position that corresponds with the greatest distance measurement as shown in the code segment that follows:

```
for turn = 1 to most_space
    gosub turn_left
next turn
```

The entire `room-mapping.bas` program is listed in Program 11.7. Compile the code and then program the PICmicro MCU 16F84 with the `room-mapping.hex` file listed in Program 11.8. Experiment with the robot's room-mapping capabilities by placing objects around the room and see where the robot chooses to walk. Another good experiment is to have the robot walk down a hallway with a few open doors. Does the robot continue to walk down the hallway or does it turn through the doorway? Analysis of the sorting algorithm will give you the answer to the robot's preferred choice when encountering two equal distances with the most space out of all the others.

After all the distance measurements have been stored in the array, another way to analyze the data would be to write an algorithm that averaged the first half of the readings and compare that value to the average of the second half of the readings. The robot could then decide to turn to the left or right, depending on the comparison.

If you wish to expand on the idea of the robot's creating and storing an internal representation of its surroundings in memory, you could use a two-dimensional array and create more complex algorithms to determine where the robot should move next. Another interesting approach is to use *certainty grids*. Hans Moravec and Alberto Elfes of Carnegie Mellon University have developed a scheme that returns certainty grids using probability distribution functions when using imprecise sonar ranging information. For each sensor reading, the assigned probability of an object being at the exact indicated range and bearing decreases radially from a maximum value at that point, according to a specified distribution function. A second distribution function characterizes the emptiness of cells between the sensor and the returned range. This technique is applied to a map where it is initially unknown if any cells are occupied.

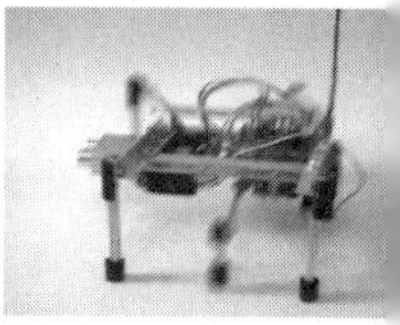

The robot is moved around the room where sonar measurements are taken and averaged to create a probability map. The robot creates its own map through exploration.

PROGRAM 11.7 Room-mapping.bas program listing.

```
'*******************************************************************
' room-mapping.bas
' The robot will take 11 distance readings from the SRF04 covering
' 180 degrees to the front of its current position. The robot will
' then move to the area with the greatest amount of free space.
' PicBasic Pro Compiler - MicroEngineering Labs
'*******************************************************************

trisa = %00000000                       ' set porta to outputs
trisb = %00001000                       ' set portb pin 3 to input

m_servo      var    byte                ' define variables and constants
l_servo      var    byte
r_servo      var    byte
timer        var    byte
right_led    var    PORTB.0
left_led     var    PORTB.1
trigger      var    PORTB.2
echo         var    PORTB.3
piezo        var    PORTB.4
right_servo  var    PORTB.5
left_servo   var    PORTB.6
mid_servo    var    PORTB.7
dist_raw     var    word
dist_inch    var    word
conv_inch    con    15                     ' conversion factor for inches
turn         var    byte
I            var    byte
best_pos     var    byte
position     var    word[12]
most_space   var    byte

low trigger                   ' set trigger pin to logic 0
low left_led                  ' turn off left LED
low right_led                 ' turn off right LED

main:

     for turn = 1 to 5    ' rotate robot 90 degrees to the left
         gosub turn_left
```

```
      next turn

      position[11] = 0              ' set last position of array to 0

      for i = 0 to 10               ' take 11 distance measurements
          gosub sr_sonar            ' get sonar reading
          position[i] = dist_inch ' store sonar reading in the array
          gosub turn_right          ' rotate the robot right by 15
                                    ' degrees
      next i                        ' do it again

      best_pos = 11

      for i = 0 to 10                  ' sorting routine to find
                                       ' location
          if position[i] >= position[best_pos] then
                best_pos = i
          endif
      next i

      most_space = 11 - best_pos        ' number of left rotations
                                        ' needed to get the
                                        ' robot back to position with
                                        ' the most space
      for turn = 1 to most_space        ' postion robot in the
                                        ' direction with the
          gosub turn_left               ' most free space.
      next turn

      for turn = 1 to 15                 ' walk forward for 15 steps
          gosub sr_sonar                 ' take sonar reading
          if dist_inch < 6 then          ' if an object is detected
then
             gosub walk_reverse           ' back the robot up a little
             goto main                    ' scan for area with the
                                          ' most space again
          endif
          Gosub walk_forward
      next turn

goto main

end

walk_forward:                             ' walk forward subroutine
    m_servo = 170
    gosub servo
```

```
    l_servo = 160
    r_servo = 160
    gosub servo
    m_servo = 100
    gosub servo
    l_servo = 120
    r_servo = 120
    gosub servo
return

walk_reverse:                               ' walk reverse subroutine
            for turn = 0 to 3
    sound portb.4,[90,1,80,2,125,1,90,2,100,2]
    m_servo = 170
    gosub servo
    l_servo = 120
    r_servo = 120
    gosub servo
    m_servo = 100
    gosub servo
    l_servo = 160
    r_servo = 160
    gosub servo
            next turn
return

turn_left:                          ' rotate the robot left 18 degrees
            sound portb.4,[140,1,80,2,125,1,95,2]
                m_servo = 170
                gosub servo
                l_servo = 120
                r_servo = 160
                gosub servo
                m_servo = 100
                gosub servo
                l_servo = 160
                r_servo = 120
                gosub servo
return

turn_right:                         ' rotate the robot right 18 degrees
            sound portb.4,[140,1,120,2,110,1,100,2]
                m_servo = 170
                gosub servo
                l_servo = 160
                r_servo = 120
```

```
                  gosub servo
                  m_servo = 100
                  gosub servo
                  l_servo = 120
                  r_servo = 160
                  gosub servo
return

sr_sonar:

         pulsout trigger,1                   ' send a 10us trigger
                                             ' pulse to the SRF04
         pulsin echo,1,dist_raw              ' start timing the
                                             ' pulse width on echo
                                             ' pin
         dist_inch = (dist_raw/conv_inch)    ' Convert raw data
                                             ' into inches
         pause 1                             ' wait for 10us before
                                             ' returning to main

return

servo:                                 ' subroutine to set servos
       for timer = 1 to 10
       pulsout mid_servo,m_servo
       pulsout left_servo,l_servo
       pulsout right_servo,r_servo
       pause 13
       next timer
return
```

PROGRAM 11.8 **Room-mapping.hex file listing.**

```
:10000000CB28A40084178004841 38E010C1C8E005E
```

```
:1000C0008C0A80300C1A8D060C198D068C188D0642
:1000D0000D0D8C0D8D0DC6288F018E00FF308E0703
:1000E000031C8F07031CC62803308D00DF307A20E5
:1000F0006E288D01E83E8C008D09FC30031C83289E
:100100008C07031880288C0764008D0F80280C183A
:1001100089288C1C8D2800008D2808008E00033053
:1001200094288E00043094289400F080D02031DBB
:100130009B280E080C0204300318013003190230 0A
:10014000140503 1DFF30C62891019001103092006 4
:100150000D0D900D910D0E0890020F08031C0F0F4E
:1001600091020318BA280E0890070F0803180F0F02
:10017000910703108C0D8D0D920BA8280C08C62832
:100180008C098D098C0A03198D0A08008313031347
:10019000831264000800831685010830860083 12EC
:1001A0000611831606118312861083168610831299
:1001B000061083160610831201 30CB00640006304F
:1001C0004B020318E6288B21CB0FDE28BE01BF01AE
:1001D000C50164000B3045020318FB28D52103102C
:1001E000450D283E840040088000840A41088000B4
:1001F000B021C50FE9280B30C400C50164000B30E5
:100200004502031823290310450D283E84000008E9
:100210009E00840A00089F000310440D283E8400BD
:10022000000008A000840A0008A1001E088C001F0816
:100230008D0021088F0020089120031D21294508E9
:10024000C400C50FFE2844080B3CC8000130CB0099
:1002500064004B084802031C30298B21CB0F28294E
:100260000130CB00640010304B0203184629D52121
:1002700040088C0041088D008F0106308E20031D40
:1002800043295A21DC284921CB0F3229DC2863007D
:100290004729AA30C700F421A030C600A030C90009
:1002A000F4216430C700F4217830C6007830C900EA
:1002B000F4210800CB01640004304B0203188A29A2
:1002C0000630A6001030A4005A308E0001303220D3
:1002D00050308E00023032207D308E0001303220CE
:1002E0005A308E000230322064308E0002303220CC
:1002F000AA30C700F4217830C6007830C900F42154
:100300006430C700F421A030C600A030C900F42139
:10031000CB0F5B2908000630A6001030A4008C30FB
:100320008E000130322050308E00023032207D307D
:100330008E00013032205F308E0002303220AA3031
:10034000C700F4217830C600A030C900F421643021
:10035000C700F421A030C6007830C900F42108009D
:100360000630A6001030A4008C308E000130322000
:1003700078308E00023032206E308E000130322014
:1003800064308E0002303220AA30C700F421A03041
```

```
:10039000C6007830C900F4216430C700F4217830F9
:1003A000C600A030C900F421080001308C008D0186
:1003B000063084000430102001308C0006308400A8
:1003C000083001200C08C2000D08C30042088C0050
:1003D00043088D000F308E008F01A420C0000D084F
:1003E000C10001306C2008000130CA0064000B30ED
:1003F0004A020318142A47088C008D010630840035
:10040000803010204608 8C008D010630840040307A
:10041000102049088C008D01063084002030102007
:0A0420000D306C20CA0FF629080009
:02400E00F53F7C
:00000001FF
```

After analyzing the `room-mapping.bas` listing, you may have noticed that the code can be optimized even more by having the distance measuring and sorting all done within the same for/next loop without using an array. The robot's internal map of its surroundings would essentially be stored in one byte of memory. This would limit the number of interesting things that could be done with the distance measurements in the array, but on a system with a very limited amount of memory, it might be desirable. Another approach for storing the distance information on a system with limited memory is to set the individual bits within a byte depending on the measurements taken. A comparison could be made that would set a bit to 0 if the distance were less than 4 feet, or set the bit to 1 if the distance were greater. The least-significant bit in the byte would be the first reading and the most-significant bit would be the last. Using this method, the robot could rotate through sixteen distance measurements and only use 2 bytes to store a rudimentary map. This method would be a cheap and fast solution. With robotics programming, there are as many different ways to code a program as there are programmers, and it all depends on time and resources.

Program 11.9 is called `mini-map.bas` and does not use an array to keep track of the sonar readings for each position. Instead, it uses two variables to store the information. The sorting of the best position is based on the greatest distance measurement and is done immediately when the sonar readings are being taken. This method uses less memory and is faster, but it does not give you the ability to analyze all the separate readings after they have been taken. So, if you wish to develop more intelligent algorithms for dealing with the sonar distance measurements, you will need to use an array (`room-mapping.bas`) or have some method to store each reading in memory. Compile `mini-map.bas` and program the 16F84 PIC with the `mini-map.hex` file listed in Program 11.10. When the robot is in operation while running `mini-map.bas`, it will function identically to the way it acted while running `room-mapping.bas`.

PROGRAM 11.9 Mini-map.bas program listing.

```
'****************************************************************
' mini-map.bas
' The robot will take 11 distance readings from the SRF04 covering
' 180 degrees but will not store the results in an array. The robot
' will then move to the area with the greatest amount of free space.
' PicBasic Pro Compiler - MicroEngineering Labs
'****************************************************************

trisa = %00000000                  ' set porta to outputs
trisb = %00001000                  ' set portb pin 3 to input

m_servo        var   byte          ' define variables and constants
l_servo        var   byte
r_servo        var   byte
timer          var   byte
right_led      var   PORTB.0
left_led       var   PORTB.1
trigger        var   PORTB.2
echo           var   PORTB.3
piezo          var   PORTB.4
right_servo    var   PORTB.5
left_servo     var   PORTB.6
mid_servo      var   PORTB.7
dist_raw       var   word
dist_inch      var   word
conv_inch      con   15             ' conversion factor for inches
turn           var   byte
i              var   byte
best_pos_index var   byte
best_pos_value var   word
most_space var byte

low trigger                  ' set trigger pin to logic 0
low left_led                 ' turn off left LED
low right_led                ' turn off right LED

main:

      for turn = 1 to 5    ' rotate robot 90 degrees to the left
          gosub turn_left
      next turn

      best_pos_index = 0        ' set initial best_pos_index to 0
      best_pos_value = 0        ' set initial best_pos_value to 0
```

```
    for i = 0 to 10
        gosub sr_sonar                        ' get sonar distance
                                              ' reading
        if dist_inch >= best_pos_value then   ' compare new
                                              ' measurement to the
                                              ' last
           best_pos_value = dist_inch         ' update best
                                              ' position if
                                              ' comparison is true
           best_pos_index = i
        endif
        gosub turn_right                      ' rotate the robot
                                              ' right by 18 degrees
    next i                                    ' do it again

    most_space = 11 - best_pos_index          ' number of left
                                              ' rotations needed
                                              ' to get the
                                              ' robot back to
                                              ' position with the
                                              ' most space
    for turn = 1 to most_space                ' postion robot in
                                              ' the direction with
                                              ' the
        gosub turn_left                       ' most free space.
    next turn

    for turn = 1 to 15                        ' walk forward for
                                              ' 15 steps
        gosub sr_sonar                        ' take sonar reading
        if dist_inch < 6 then                 ' if an object is
                                              ' detected then
           gosub walk_reverse                 ' back the robot up
                                              ' a little
           goto main                          ' scan for area with
                                              ' the most space
                                              ' again
        endif
        Gosub walk_forward
    next turn

goto main

end

walk_forward:                          ' walk forward subroutine
   m_servo = 170
```

```
    gosub servo
    l_servo = 160
    r_servo = 160
    gosub servo
    m_servo = 100
    gosub servo
    l_servo = 120
    r_servo = 120
    gosub servo
return

walk_reverse:                               ' walk reverse subroutine
               for turn = 0 to 3
    sound portb.4,[90,1,80,2,125,1,90,2,100,2]
    m_servo = 170
    gosub servo
    l_servo = 120
    r_servo = 120
    gosub servo
    m_servo = 100
    gosub servo
    l_servo = 160
    r_servo = 160
    gosub servo
               next turn
return

turn_left:                           ' rotate the robot left 18 degrees
               sound portb.4,[140,1,80,2,125,1,95,2]
                    m_servo = 170
                    gosub servo
                    l_servo = 120
                    r_servo = 160
                    gosub servo
                    m_servo = 100
                    gosub servo
                    l_servo = 160
                    r_servo = 120
                    gosub servo
return

turn_right:                          ' rotate the robot right 18 degrees
               sound portb.4,[140,1,120,2,110,1,100,2]
                    m_servo = 170
                    gosub servo
                    l_servo = 160
                    r_servo = 120
```

```
                    gosub servo
                    m_servo = 100
                    gosub servo
                    l_servo = 120
                    r_servo = 160
                    gosub servo
return

sr_sonar:

        pulsout trigger,1                ' send a 10us trigger
                                         ' pulse to the SRF04
        pulsin echo,1,dist_raw           ' start timing the
                                         ' pulse width on echo
                                         ' pin
        dist_inch = (dist_raw/conv_inch) ' Convert raw data
                                         ' into inches
        pause 1                          ' wait for 10us before
                                         ' returning to main

return

servo:                                   ' subroutine to set servos
        for timer = 1 to 10
        pulsout mid_servo,m_servo
        pulsout left_servo,l_servo
        pulsout right_servo,r_servo
        pause 13
        next timer
return
```

PROGRAM 11.10   Mini-map.hex file listing.

```
:10000000CB28A0008417800484138E010C1C8E0062
:1000100023200319C62823200319C6282320C62815
:10002000A00059200C080D040319C628C020841311
:100030002008800664001C281D288C0A03198D0FD7
:100040001A288006C62820088E0601308C008D01F3
:100050000000820050E06031D08008C0A03198D0FE9
:1000600028280800 8F00220884 0020095A208413C1
:100070008F080319C628F03091000E0880389000D0
:10008000F0309103031991000 3198F030319C62857
:100090 0049285D2003010C1820088E1F20088E08B7
:1000A0000 03190301900F562880063D2857280000A9
:1000B0004028FF3A84178005C6280D080C04031950
:1000C0008C0A80300C1A8D060C198D068C188D0642
```

```
:1000D0000D0D8C0D8D0DC6288F018E00FF308E0703
:1000E000031C8F07031CC62803308D00DF307A20E5
:1000F0006E288D01E83E8C008D09FC30031C83289E
:100100008C07031880288C0764008D0F80280C183A
:1001100089288C1C8D2800008D2808008E00033053
:1001200094288E00043094289400 0F080D02031DBB
:100130009B280E080C0204300318013003190230 0A
:10014000140503 1DFF30C62891019001103092 0064
:100150000D0D900D910D0E0890020F08031C0F0F4E
:1001600091020318BA280E0890070F0803180F0F02
:1001700091070310 8C0D8D0D920BA8280C08C62832
:100180008C098D098C0A03198D0A08008313031347
:1001900083126400080083168501083086008312EC
:1001A000061183160611831286108316861083129 9
:1001B000061083160610831201 30B1006400063069
:1001C00031020318E6286B21B10FDE28AA01A40131
:1001D000A501AB0164000B302B0203180329B521E4
:1001E00026088C0027088D0025088F002408912000
:1001F000031D00292608A4002708A5002B08AA0033
:1002000009021AB0FEA282A080B3CAE000130B10068
:10021000640031082E02031C10296B21B10F08293C
:100220000130B10064001030310203182629B521D5
:1002300026088C0027088D008F0106308E20031DB4
:1002400023293A21DC282921B10F1229DC28630057
:100250002729AA30AD00D421A030AC00A030AF00D7
:10026000D4216430AD00D4217830AC007830AF00B8
:10027000D4210800B10164000430310203186A2956
:100280000630A2001030A0005A308E00013032201B
:1002900050308E00023032207D308E00013032200E
:1002A0005A308E00023032206430 8E00023032200C
:1002B000AA30AD00D4217830AC007830AF00D42122
:1002C0006430AD00D421A030AC00A030AF00D42108
:1002D000B10F3B2908000630A2001030A0008C307E
:1002E0008E000130322050308E00023032207D30BE
:1002F0008E00013032205F308E0002303220AA3072
:10030000AD00D4217830AC00A030AF00D4216430EF
:10031000AD00D421A030AC007830AF00D42108006B
:100320000630A2001030A0008C308E000130322048
:1003300078308E00023032206E308E000130322054
:1003400064308E0002303220AA30AD00D421A030BB
:10035000AC007830AF00D4216430AD00D4217830C7
:10036000AC00A030AF00D421080001308C008D011A
:100370000630840004301020013 08C0006308400E8
:1003800083001200C08A8000D08A90028088C00DE
:1003900029088D000F308E008F01A420A6000D08C3
:1003A000A70001306C2008000130B00064000B3061
:1003B00030020318F4292D088C008D0106308400CA
```

```
:1003C000803010202C088C008D01063084004030D5
:1003D00010202F088C008D01063084002030102062
:0A03E0000D306C20B00FD629080084
:02400E00F53F7C
:00000001FF
```

Maze Solving

A popular activity among robot experimenters is robotic maze solving. The objective is to use a robot to solve a maze as quickly as possible. The robot does not know the configuration of the maze before its first run-through. When looking down on a maze, you may think that it is easy to find the shortest path from the start to the finish. But from the robot's perspective, the problem is not quite so easy. Most hobbyist robots can only sense whether there is a wall directly ahead, to the left or to the right. Based on this limited information, the robot must use proven algorithms to explore and find its way from the start to the finish. A good maze-solving robot will try to find the shortest path to the finish destination in order to get there as fast as possible.

There are a few popular methods used in maze solving. Wall following is the simplest method used most often, but it will not solve mazes that contain islands. For the following experiment, we will assume that the maze has connecting walls and does not contain islands. Figure 11.16 shows a simple maze that still contains dead-end areas where the robot could potentially get stuck.

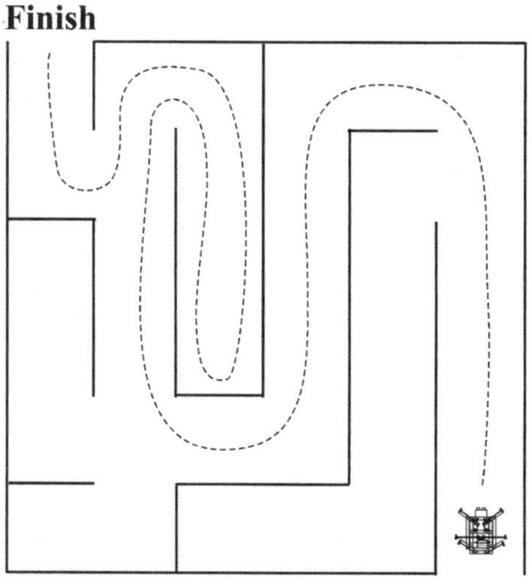

FIGURE 11.16 A simple maze with dead ends.

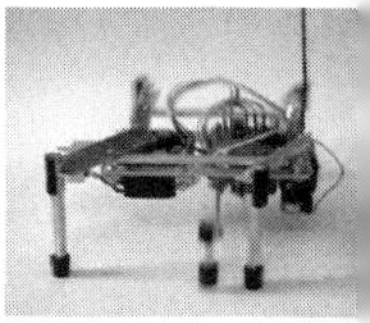

Using the wall-following method, the robot finds either the right or left wall and follows it until it has reached the finish point. With this method, the robot will never go back over paths it has already tried. The advantage to this is that the robot does not have to create an entire map of the maze in memory before trying to solve the maze, or remember where it has already been. Sometimes this method solves mazes faster than methods that map, analyze, and then guide the robot through the maze because the robot starts traveling through the maze right away and solves as it goes. The steps to solving mazes using the wall-following method for mazes like the one shown in Figure 11.16 are as follows:

1. If there is not a wall on the right hand side of the robot, then turn the robot to the right.
2. If there is not a wall to the front of the robot, then move the robot one cell forward.
3. If steps 1 and 2 are not executed, then turn the robot to the left.
4. Continue steps 1 through 3 until the robot is at the finish destination, then stop.

When the robot encounters a dead-end situation using this scheme to solve a maze, it just turns itself around, according to the rules, and backtracks its way out as shown by the robot's walking path in Figure 11.16.

The wall-following maze-solving method has been implemented in Program 11.11, called `maze-solving.bas`. Compile `maze-solving.bas` and program the PICmicro MCU 16F84 with the `maze-solving.hex` file listed in Program 11.12. Set up a small maze with a start and a finish. Leave about 2 feet between walls. Let the robot try to solve the maze. You might notice a problem when the robot is solving the maze. The problem with walking robots is that, when moving forward, they do not always walk in a straight line. Most maze-solving robots use wheeled bases that are aligned so that the machine moves forward quickly in a straight line. Quite often the walking robot wanders to the left or right. Over time, if these small discrepancies are not corrected, the robot may be off by 90 degrees from its original orientation. Since this algorithm is counting on the robot's being correctly oriented each time it makes a turn, this ends up being a problem for us. This could easily be corrected by interfacing one of the electronic compass devices or gyroscopes that are available on the market. When the robot needs to turn to the right, the information from the compass could be used to make sure that the robot has rotated completely from an orientation of north to east, for example.

PROGRAM 11.11 Maze-solving.bas program listing.

```
'*************************************************************
' maze-solving.bas
' Walking robot maze solving using wall following technique.
' PicBasic Pro Compiler
'*************************************************************

trisa = %00000000               ' set porta to outputs
trisb = %00001000               ' set portb pin 3 to input

m_servo      var   byte         ' define variables and constants
l_servo      var   byte
r_servo      var   byte
timer        var   byte
right_led    var   PORTB.0
left_led     var   PORTB.1
trigger      var   PORTB.2
echo         var   PORTB.3
piezo        var   PORTB.4
right_servo  var   PORTB.5
left_servo   var   PORTB.6
mid_servo    var   PORTB.7
dist_raw     var   word
dist_inch    var   word
conv_inch    con   15           ' conversion factor for inches
turn         var   byte

low trigger                     ' set trigger pin to logic 0
low left_led                    ' turn off left LED
low right_led                   ' turn off right LED

main:

     for turn = 1 to 5          ' look to the right
     gosub turn_right
     next turn

     gosub sr_sonar             ' get sonar distance measurement

     if dist_inch < 7 then   ' is there a wall to the right?
         for turn = 1 to 5   ' Yes, then turn the robot back
         gosub turn_left     ' No, then keep the robot facing right
         next turn
     endif
```

```
      gosub sr_sonar                ' get sonar distance measurement

      if dist_inch < 7 then         ' is there a wall in front of the
                                    ' robot?
         for turn = 1 to 5          ' Yes, then turn the robot left
         gosub turn_left            ' No, keep the robot in its current
                                    ' orientation
         next turn
      endif

      for turn = 1 to 5             ' walk the robot forward 5 cycles
      gosub walk_forward
      next turn

goto main

end

walk_forward:                       ' forward walking subroutine
     m_servo = 170
     gosub servo
     l_servo = 160
     r_servo = 160
     gosub servo
     m_servo = 100
     gosub servo
     l_servo = 120
     r_servo = 120
     gosub servo
return

walk_reverse:                       ' walk reverse subroutine
     sound portb.4,[90,1,80,2,125,1,90,2,100,2]
     m_servo = 170
     gosub servo
     l_servo = 120
     r_servo = 120
     gosub servo
     m_servo = 100
     gosub servo
     l_servo = 160
     r_servo = 160
     gosub servo
return

turn_left:                    ' rotate the robot left 18 degrees
            sound portb.4,[140,1,80,2,125,1,95,2]
```

```
                    m_servo = 170
                    gosub servo
                    l_servo = 120
                    r_servo = 160
                    gosub servo
                    m_servo = 100
                    gosub servo
                    l_servo = 160
                    r_servo = 120
                    gosub servo
return

turn_right:                         ' rotate the robot right 18 degrees
               sound portb.4,[140,1,120,2,110,1,100,2]
                    m_servo = 170
                    gosub servo
                    l_servo = 160
                    r_servo = 120
                    gosub servo
                    m_servo = 100
                    gosub servo
                    l_servo = 120
                    r_servo = 160
                    gosub servo
return

sr_sonar:

          pulsout trigger,1                    ' send a 10us trigger
                                               ' pulse to the SRF04
          pulsin echo,1,dist_raw               ' start timing the
                                               ' pulse width on echo
                                               ' pin
          dist_inch = (dist_raw/conv_inch)     ' Convert raw data
                                               ' into inches
          pause 1                              ' wait for 10us before
                                               ' returning to main

return

servo:                                    ' subroutine to set servos
        for timer = 1 to 10
        pulsout mid_servo,m_servo
        pulsout left_servo,l_servo
        pulsout right_servo,r_servo
        pause 13
        next timer
return
```

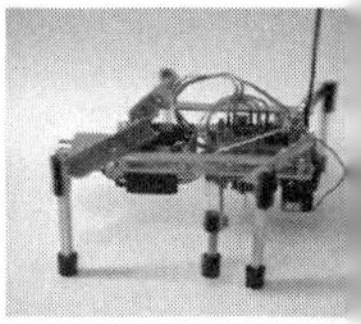

PROGRAM 11.12 Maze-solving.hex file listing.

```
:10000000C828A000841780048413 8E010C1C8E0065
:1000100023200319C32823200319C3282320C3281E
:10002000A00059200C080D040319C328BD20841317
:100030002008800664001C281D288C0A03198D0FD7
:100040001A288006C32820088E0601308C008D01F6
:1000500000820050E06031D08008C0A03198D0FE9
:10006000282808008F002208840020095A208413C1
:100070008F080319C328F03091000E0880389000D3
:10008000F030910303199100031 98F030319C3285A
:1000900049285D2003010C1820088E1F20088E08B7
:1000A0000319030190 0F562880063D2857280000A9
:1000B0004028FF3A84178005C3280D080C04031953
:1000C0008C0A80300C1A8D060C198D068C188D0642
:1000D0000D0D8C0D8D0DC3288F018E00FF308E0706
:1000E000031C8F07031CC32803308D00DF307A20E8
:1000F0006E288D01E83E8C008D09FC30031C83289E
:100100008C07031880288C0764008D0F80280C183A
:1001100089288C1C8D2800008D2808008E00033053
:10012000912894000F080D02031D98280E080C0258
:10013000043003180130031902301405031DFF3089
:10014000C3289101900110309200 0D0D900D910D7A
:100150000E0890020F08031C0F0F91020318B72816
:100160000E0890070F0803180F0F910703108C0D4E
:100170008D0D920BA5280C08C3288C098D098C0ABB
:1001800003198D0A0800831303138312640008 0007
:1001900083168501083086008312061183160611 26
:1001A00083120610831606108312861083168610 9B
:1001B00083120130AC00640006302C020318E328DF
:1001C0007721AC0FDB289C2124088C0025088D00AA
:1001D0008F0107308E20031DF7280130AC0064002A
:1001E00006302C020318F7285221AC0FEF289C216F
:1001F00024088C0025088D008F0107308E20031DF8
:100200000B290130AC00640006302C0203180B29C6
:100210005221AC0F03290130AC00640006302C02DF
:1002200003181529 1821AC0F0D29D92863001629A8
:10023000AA30A900BB21A030A800A030AA00BB2191
:100240006430A900BB217830A8007830AA00BB2117
:1002500008000630A2001030A0005A308E00013095
:10026000322050308E00023032207D308E0001303E
:1002700032205A308E00023032206430 8E0002303C
:100280003220AA30A900BB217830A8007830AA001B
:10029000BB216430A900BB21A030A800A030AA0077
:1002A000BB2108000630A2001030A0008C308E0068
:1002B00001303220 50308E00023032207D308E00EE
:1002C000013032205F308E0002303220AA30A90087
```

```
:1002D000BB217830A800A030AA00BB216430A9005F
:1002E000BB21A030A8007830AA00BB21080006304E
:1002F000A2001030A0008C308E000130322078300 7
:100300008E00023032206E308E0001303220643098
:100310008E0002303220AA30A900BB21A030A800F4
:100320007830AA00BB216430A900BB217830A80036
:10033000A030AA00BB21080001308C008D010630DE
:100340008400043010200130 8C0006308400083016
:1003500001200C08A6000D08A70026088C0027081D
:100360008D000F308E008F01A120A4000D08A50084
:1003700001306C2008000130AB0064000B302B0210
:100380000318DB2929088C008D0106308400803099
:1003900010202808 8C008D0106308400403010 2089
:1003A0002A088C008D01063084002030102000D308A
:0803B0006C20AB0FBD29080011
:02400E00F53F7C
:00000001FF
```

Summary

The addition of the sonar ranger gives the robot the ability to sense objects at much greater distances and accuracy than if we use the infrared method discussed earlier. A further enhancement would be to mount the sonar module on a servo, and have the robot scan the area without having to stop to rotate its entire body. The robot would then have the ability to continue moving through an area while scanning for objects at the same time.

In Chapter 12, we will explore remotely controlling the robot by implementing a radio transmitter and receiver. The sonar sensor will still be used to signal to the radio operator any impending obstacles or to inhibit forward walking if an obstacle is present.

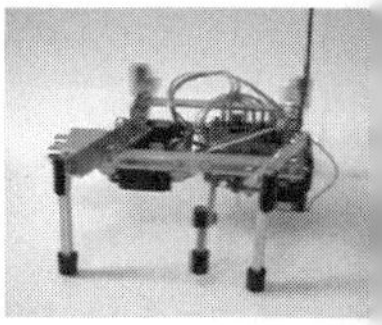

12

RADIO REMOTE CONTROL

Up until now, the robot has been operating autonomously under its own cognition. This chapter will deal with interfacing a model aircraft transmitter and receiver for human control of the robot. The benefit to using a long-range aircraft remote control (R/C) is that the robot can be guided to exact remote locations, and other functions such as a gripper or video camera can be added and controlled. A remote-controlled walking robot has many uses, such as locating survivors in disaster areas, safe land mine search and removal, military reconnaissance, and espionage applications. The model airplane remote control system that will be used to control the robot is called the Quattro. This system is made up of the NET-E104 transmitter and the NER-700 receiver as shown in Figure 12.1. This system is manufactured by JR Remote Control and is available at most hobby R/C airplane shops. Horizon Hobby Inc. (http://www.horizonhobby.com/) distributes this and other remote control systems. It has a transmitting distance of 1 mile and operates in the 72-MHz range. Transmitters within the 72-MHz range do not require a special license to operate. The instructions given to interface the transmitter and receiver will work with any standard R/C system and are not specific to this make and model. The parts required to add the remote control are listed in Table 12.1.

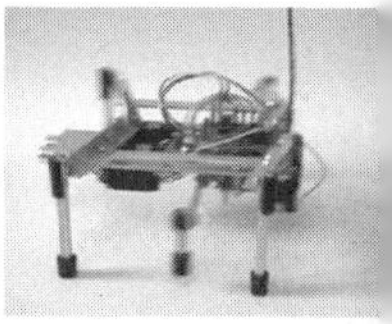

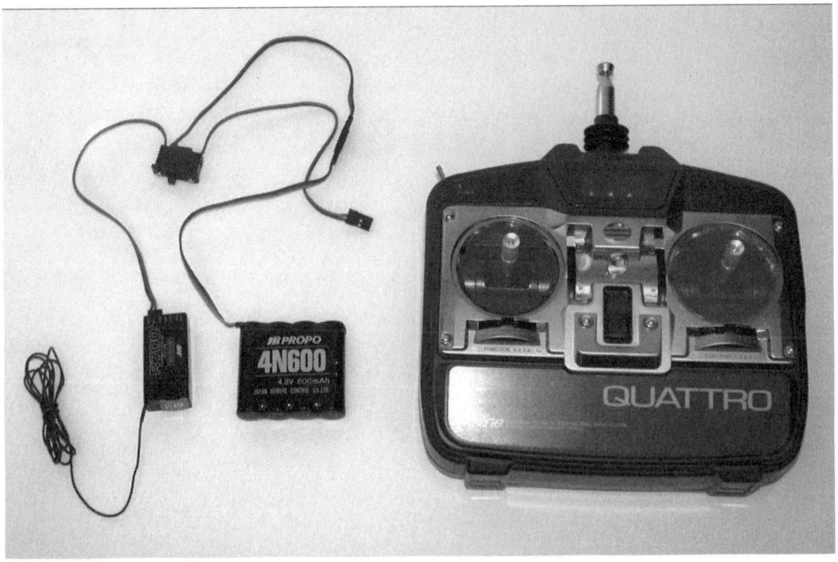

FIGURE 12.1 Quattro airplane remote control transmitter and R700 receiver.

TABLE 12.1 Parts List for the Addition of a Remote Control System

PART	QUANTITY	DESCRIPTION
NET-E104 transmitter	1	JR remote control product
NER-700 receiver	1	JR remote control product
Servo leads	3	Leads with a single connector
3-connector female header	2	2.5-mm spacing
2-connector female header	1	2.5-mm spacing
Heat-shrink tubing	4.5 inches	1/8-inch diameter
# 24 gauge wire	8 1/2 inches	Rigid and bendable

To connect the remote control receiver shown in Figure 12.2 to the robot, it will be necessary to fabricate three servo lead wires. For this application, it is recommended that you buy three servo leads with a servo connector on one end. Servo leads like the one shown in Figure 12.3 can be purchased from a local R/C hobby shop or servo supplier. There are two reasons to buy the servo lead with the servo connector. The first is that the connectors on the receiver have rounded edges to ensure that the connector is not plugged in backwards. The second is that the connectors on the receiver are positioned close together and the connectors we fabricate might be bulky and may not fit properly.

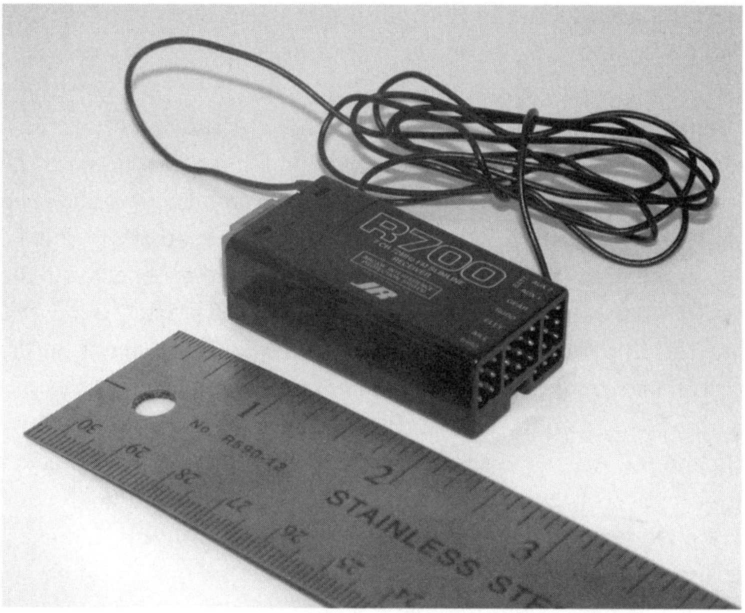

FIGURE 12.2 The JR NER-700 radio receiver—lightweight and small.

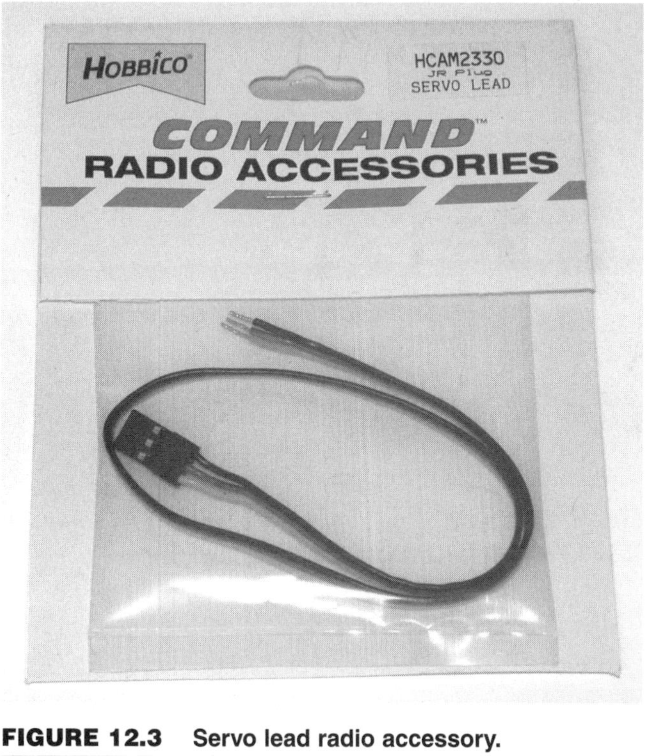

FIGURE 12.3 Servo lead radio accessory.

Cut nine pieces of heat-shrink tubing to a length of 3/8 of an inch. The first two connectors that need to be fabricated will connect two of the radio receiver channels to the robot's controller board. Take one of the servo leads and slide one piece of the 3/8-inch heat-shrink tubing over each of the three bare wires and past where the plastic has been stripped from the wire. Solder a 3-connector female header onto the stripped wires. Push the heat-shrink tubing up over the solder connections and then shrink the tubing tight with a cigarette lighter or heat gun. Make another servo lead, following the previous procedure, for a total of two identical servo connectors. The third connector is similar to the other two but will only have the +5V and ground (red and black) wires connected to power the radio receiver from the 5V regulator on the robot's controller board. Take the third servo lead and cut the stripped end of the yellow wire off at a length of 1 inch from the end. Place a piece of heat-shrink tubing over the yellow, black, and red wires and shrink in place. Slide the remaining pieces of heat-shrink tubing over each of the two remaining bare wires (red and black) and solder a 2-connector female header onto the wires. Push the heat-shrink tubing up over the solder connectors and shrink in place with a heat source. Figure 12.4 shows what the finished connector ends should resemble.

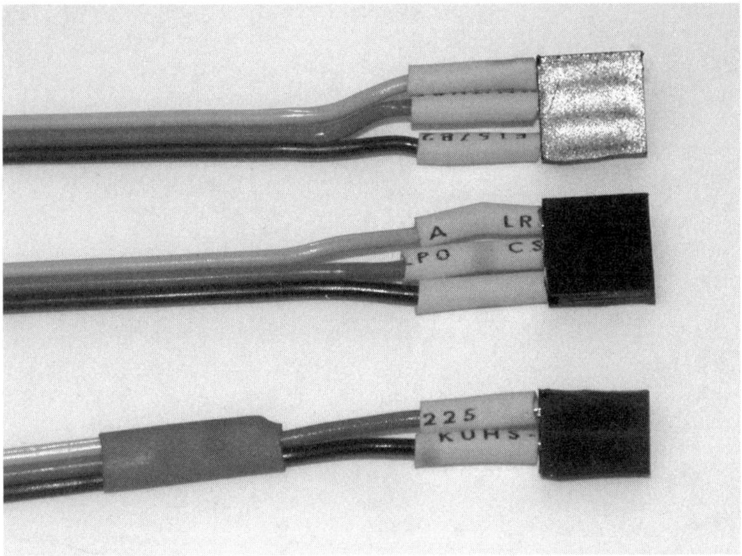

FIGURE 12.4 **Finished radio receiver connector leads.**

Use Figure 12.5 as a guide to connect the radio receiver to the robot's main controller board with the three cables that were just fabricated. Make sure when you connect the power and ground from the receiver to the controller board that the red wire connects to the +5V power connection and the black goes to ground. The cable that connects to JP1 on the controller board is connected to the channel on the receiver marked as "ELEV" (elevator) and JP2 is connected to the channel marked "AILE" (aileron). The ends of the cables with the female connectors added are attached to the robot controller board and the servo connectors that were already attached to the cables plug into the receiver. The elevator and aileron channels on airplane remote controls are most often mapped to the right-hand control stick of the remote control transmitter. We will be using the right-hand control stick to give the robot walking motion commands.

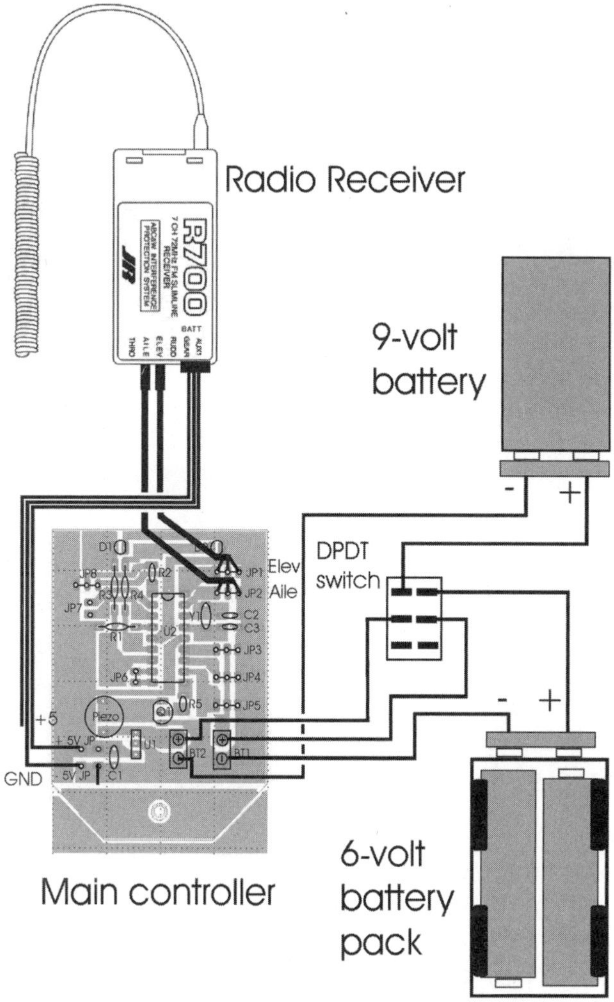

FIGURE 12.5 Radio receiver connection diagram.

Mount the radio receiver to the robot using a small Velcro® fastener or double-sided tape. Stick one side of the Velcro or tape to the back of the receiver and the other to the bottom of the 6V battery pack holder as shown in Figure 12.6. The next step is to create a post on which to mount the antenna. Use a bendable yet rigid piece of #24 gauge wire; cut to a length of 8 1/2 inches. Make a small loop at the end of the wire and then bend it on a 90-degree angle. Mount the antenna post to the robot using the existing machine screw and nut that is holding the 6V battery pack and controller circuit board to the robot. Tightly wind the antenna wire around the antenna post all the way up to the top as shown in Figures 12.6 and 12.7.

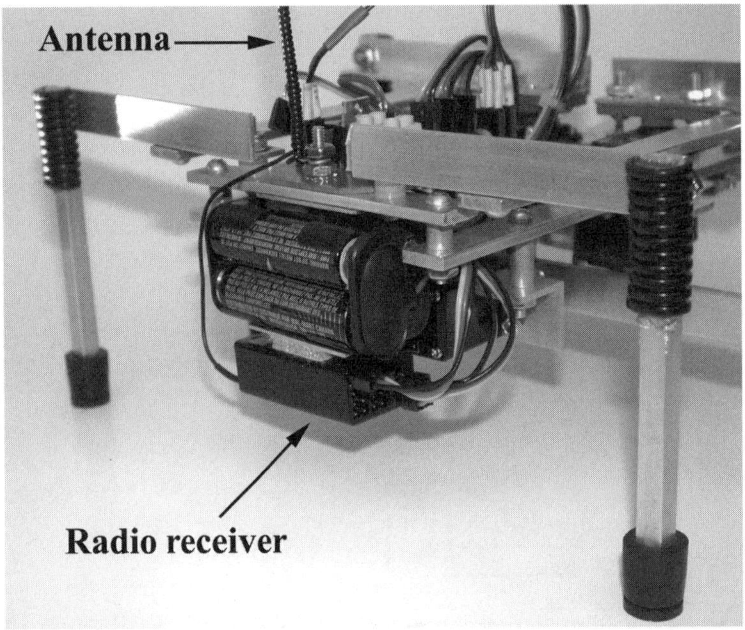

FIGURE 12.6 Radio receiver and antenna attached to the robot—back view.

Interpreting Signals from the Radio Receiver

The first experiment will be to analyze the signals from the radio receiver and show the information on the LCD display. This information will be needed when developing the remote control software. Temporarily disconnect the power to the ultrasonic sensor module on the controller board and use that connection to power the LCD dis-

FIGURE 12.7 Robot with antenna attached—front view.

play. The serial connection for the LCD display should be attached to JP7 as shown in Figure 11.12 in Chapter 11.

The signals produced by the radio receiver are exactly the same as the signals we have been creating with our software to control the servos. As was explained in Chapter 10, the signal is a positive square wave with a pulse width between 1 and 2 milliseconds, and a frequency between 50 and 60 hertz. To analyze the signals, we will use the PicBasic `pulsin` command to measure the width of the pulses on each of the two channels and then display the results. The channel on the receiver marked as "AILE" is connected to the PICmicro MCU 16F84 PORTA.0 pin and the channel marked "ELEV" is connected to the PORTA.1 pin. We will refer to the aileron channel as channel 1. It will control whether the robot turns left or right. The elevator channel will be referred to as channel 2, and will control the forward and reverse

movements of the robot. There will be eight different movements based on the position of the control stick, which determines the interaction between channel 1 and channel 2. Figure 12.8 is a diagram of the radio transmitter showing the desired robot movements that correspond to the positions of the control stick. There are nine separate positions of the control stick and nine corresponding actions that the robot will take. The robot will either remain motionless, walk forward, reverse, turn left on the spot, turn right on the spot, turn right while walking forward, turn right while walking in reverse, turn left while walking forward, or turn left while walking in reverse.

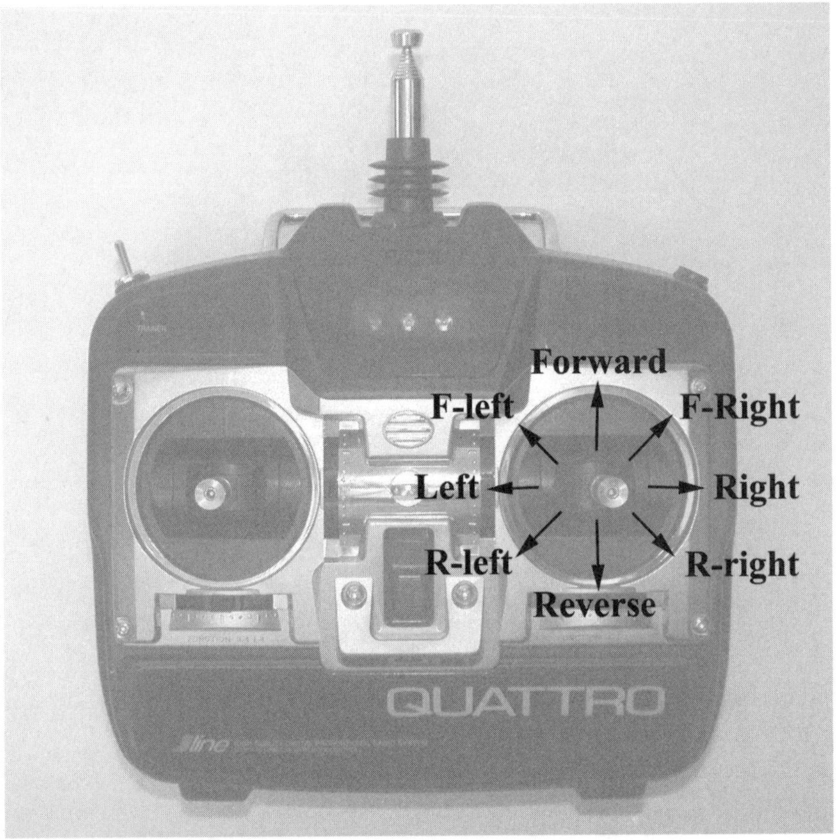

FIGURE 12.8 **Radio transmitter showing control stick positions and robot actions.**

The program to analyze the two radio receiver channels is listed in Program 12.1 and is named `radio-analysis.bas`. Compile `radio-analysis.bas` and then program the PICmicro MCU 16F84 with the `radio-analysis.hex` file listed in Program 12.2.

PROGRAM 12.1 radio-analysis.bas program listing.

```
'*******************************************************************
' radio-analysis.bas
' Program to measure the pulse widths of 2 channels from the R700
' radio receiver then display the results on the LCD display.
'*******************************************************************

Include "modedefs.bas"              ' Include serial modes

trisa = %00000011                   ' set porta pins 0&1 to inputs
trisb = %00001000                   ' set portb pin 3 to input

right_led  var      PORTB.0         ' define variables and constants
left_led   var      PORTB.1
lcd        var      PORTA.3
piezo      var      PORTB.4
chan_1     var      PORTA.0
chan_2     var      PORTA.1
baud       con      N2400
Channel_1  var      byte
Channel_2  var      byte

low left_led                        ' turn off left LED
low right_led                       ' turn off right LED

Serout lcd,baud,[254,1]             ' clear lcd screen

pause 1000                          ' give time to initialize lcd

Serout lcd,baud,[254,128,"Radio Control"]  ' display program title
                                           ' on the LCD

sound piezo,[100,10,50,5,70,10,50,2]       ' Make startup sound

pause 1000                                 ' wait for 1 second

Serout lcd,baud,[254,128,"channel-1:  "]   ' set up the LCD display

Serout lcd,baud,[254,192,"Channel-2:  "]

main:

     pulsin chan_1,1,channel_1
     pulsin chan_2,1,channel_2
```

```
    Serout lcd,baud,[254,138,#channel_1,"   "]        ' display
                                                       ' channel 1
                                                       ' pulse width
                                                       ' value


    Serout lcd,baud,[254,202,#channel_2,"   "]        ' display
                                                       ' channel 2
                                                       ' pulse width
                                                       ' value


Goto main


End
```

PROGRAM 12.2   **Radio-analysis.hex file listing.**

```
:10000000F528A0008417800484138E010C1C8E0038
:1000100010200319F02810200319F0281020F028D0
:1000200020088E0601308C008D01000820050E0688
:10003000031D08008C0A03198D0F15280800910173
:1000400009009F1727308F001030322003308F0030
:10005000E83032208F01643032208F010A303220A4
:1000600010083D288E0011088D0010088C00D42047
:100070000C08031D9F139F1B0800303E92002208AE
:100080008400093093000310 4A20920C930B4428FB
:1000900003144A2884139F1D5928000820041F1D9B
:1000A00020068000841700082004031C200680001E
:1000B0006428000820040 31C20061F192006800065
:1000C000084172009800564281F0D06398C006D20D7
:1000D0008D008C0A6D200000C02800308A000C08BA
:1000E00082070134753403341534 00343C340C3445
:1000F000D9348F00220884002009A02084138F089F
:1001000000319F028F03091000E0880389000F0308C
:100110009103031991000319 8F030319F028902804
:10012000A32003010C1820088E1F20088E08031935
:1001300000301900F9D28800684289E2800008728B0
:1001400084178005F0280D080C0403198C0A8030F0
:10015000C1A8D060C198D068C188D060D0D8C0D44
:100160008D0DF0288F018E00FF308E07031C8F0746
:10017000031CF02803308D00DF30C020B4288D012F
:10018000E83E8C008D09FC30031CC9288C0703183D
:10019000C6288C0764008D0FC6280C18CF288C1C2D
:1001A000D3280000D3280800910190011030920 05C
:1001B0000D0D900D910D0E0890020F08031C0F0FEE
:1001C00091020318EA280E0890070F0803180F0F72
:1001D000910703108C0D8D0D920BD8280C08F02878
```

```
:1001E0008313031383126400080083160330850011
:1001F00008308600831206108316061083128610BC
:10020000831686100530830312A2000830A000043047
:100210009F00FE303E2001303E2003308F00E8304A
:10022000B3200530A2000830A00004309F00FE304B
:100230003E2080303E2052303E2061303E206430EF
:100240003E2069303E206F303E2020303E2043303B
:100250003E206F303E206E303E2074303E207230A3
:100260003E206F303E206C303E200630A200103021
:10027000A00064308E000A30792032308E000530C4
:100280000792046308E000A30792032308E000230DC
:100290000792003308F00E830B3200530A200083009
:1002A000A00004309F00FE303E2080303E206330AE
:1002B0003E2068303E2061303E206E303E206E3061
:1002C0003E2065303E206C303E202D303E203130C7
:1002D0003E203A303E2020303E2020303E2020304C
:1002E0003E2020303E2020303E2020303E20053071
:1002F000A2000830A00004309F00FE303E20C03035
:100300003E2043303E2068303E2061303E206E303B
:100310003E206E303E2065303E206C303E202D3039
:100320003E2032303E203A303E2020303E202030E9
:100330003E2020303E2020303E2020303E20203005
:100340003E2001308C0005308400013001200C0873
:10035000A40001308C0005308400023001200C081C
:10036000A5000530A2000830A00004309F00FE3038
:100370003E208A303E2024081F2020303E2020309E
:100380003E2020303E200530A2000830A00004307E
:100390009F00FE303E20CA303E2025081F2020301E
:0C03A0003E2020303E2020303E20A129CD
:02400E00F53F7C
:00000001FF
```

After the PIC has been inserted into the 18-pin socket on the controller board, turn on the radio transmitter and then power up the robot. When the program is running, the pulse width values for channels 1 and 2 should be displayed on the LCD. These values should be between 140 and 155 when the stick is not moved. The typical value when the stick is in the home middle position is 150. Move the control stick slowly towards the forward position as shown in Figure 12.8. The value displayed on the LCD for channel 2 will increase incrementally from 150 to 190 when the stick is moved as far forward as it can go. Move the control stick all the way down to the reverse position. The displayed value for channel 2 will decrease in value from 150 in the middle position to 105 at the end of its travel. When the control is moved all the way to the right, the maximum value for channel 1 is 105. When the control is moved to the furthest left position the value for channel 1 is 190. The combination of these values will be used in the robot control program that will be developed later.

Remote Control

The first remote control experiment will be to turn on the robot's light-emitting diodes and produce a tone when the control stick is moved into the four main positions of forward, reverse, right, and left. Based on the values obtained by analyzing channels 1 and 2 in the last experiment, thresholds can be set that will ensure that the required actions are taken. If single values are hard-coded into the program and the receiver's output signals drift a little, then the desired action will not be taken. It is good to add a slight buffer zone to the boundaries by padding the values so that any small inconsistencies in the signal are ignored. Compile `led-remote.bas` listed in Program 12.3. Program the PICmicro MCU 16F84 with the `led-remote.hex` file listed in Program 12.4 and then insert it into the 18-pin socket on the controller board. The comparison values are set so that the program will react to the input signal when it gets above or below the threshold levels with a certain amount of padding built in to make sure that a window of values will work as shown in the program listing. Turn on the transmitter and then turn on the robot. Try out the different control positions and see the reaction with the light-emitting diodes and the sound produced. Try turning off the transmitter with the robot still powered up and see what happens. Because the program has not taken into consideration the situations where the transmitter power is off or out of range, the LEDs will turn on and off randomly.

PROGRAM 12.3 Led-remote.bas program listing.

```
'***********************************************************************
' led-remote.bas
' Turn on left and right LEDs depending on the input values from
' the radio receiver.
' Forward - right LED on
' Reverse - left LED on
' Right - both LEDs on
' Left - flash both LEDs
' PicBasic Pro compiler - MicroEngineering Labs.
'***********************************************************************

trisa = %00000011                  ' set porta pins 0&1 to inputs
trisb = %00001000                  ' set portb pin 3 to input

                                   ' define variables and constants

piezo       var      PORTB.4
right_led   var      PORTB.0
left_led    var      PORTB.1
```

```
chan_1      var     PORTA.0
chan_2      var     PORTA.1
Channel_1   var     byte
Channel_2   var     byte
I           var     byte

low left_led                            ' turn off left LED
low right_led                           ' turn off right LED

sound piezo,[100,10,50,5,70,10,50,2]    ' Make startup sound

main:

     pulsin chan_1,1,channel_1
     pulsin chan_2,1,channel_2

     If channel_2 > 180 then
        sound piezo,[100,10,50,5]        ' Make sound
        high right_led                   ' turn on right LED
        low left_led
     endif

     If channel_2 < 120 then
        sound piezo,[50,5,100,10,30,20]  ' Make sound
        low right_led
        high left_led                    'turn on left LED
     endif

     If channel_1 < 120 then
        sound piezo,[115,10,80,5,50,20]  ' Make sound
        high right_led                   ' Turn on both LEDs
        high left_led
     endif

     If channel_1 > 180 then
        sound piezo,[50,20,80,5,115,10]  ' Make sound
        for I = 1 to 7                   ' flash both LEDs
            high right_led
            high left_led
            pause 100
            low right_led
            low left_led
            pause 100
        next I
     endif
```

```
Goto main

End
```

PROGRAM 12.4   Led-remote.hex file listing.

```
:100000007F28A0008417800484138E010C1C8E00AE
:10001000102003197A28102003197A2810207A2832
:1000200020088E0601308C008D01000820050E0688
:10003000031D08008C0A03198D0F152808008F0076
:100040002208840020094620841 38F0803197A2887
:10005000F03091000E0880389000F03091030319C1
:10006000910003198F0303197A28362849200301C8
:100070000C1820088E1F20088E0803190301900F0A
:10008000432880062A28442800002D28841780054C
:100090007A280D080C0403198C0A80300C1A8D067E
:1000A0000C198D068C188D060D0D8C0D8D0D7A2872
:1000B0008F018E00FF308E07031C8F07031C7A28E8
:1000C00003308D00DF3066205A288D01E83E8C0019
:1000D0008D09FC30031C6F288C0703186C288C07D3
:1000E00064008D0F6C280C1875288C1C7928000072
:1000F00079280800831303138312640008008316 11
:10010000033085000830860083128610831686101F
:10011000831206108316061006308312A2001030D8
:10012000A00064308E000A301F2032308E0005306F
:100130001F2046308E000A301F2032308E000230E1
:100140001F2001308C0005308400013001200C0894
:10015000A40001308C0005308400023001200C081E
:10016000A5006400B5302502031CCA280630A20091
:100170001030A00064308E000A301F2032308E0014
:1001800005301F200614831606108312861083166E
:10019000861083126400783025020318E7280630A1
:1001A000A2001030A00032308E0005301F206430D5
:1001B0008E000A301F201E308E0014301F200610C3
:1001C0008316061083128614831686108312640029
:1001D000783024020318042906 30A2001030A00051
:1001E00073308E000A301F2050308E0005301F20E3
:1001F00032308E0014301F200614831606108312 2E
:100200008614831686108312640 0B5302402031C02
:1002100036290630A2001030A00032308E001430 93
:100220001F2050308E0005301F2073308E000A30A2
:100230001F200130A60064000830260203183629 6A
:10024000061483160610831286148316861064 30F3
:10025000831258200610831606108312861083160 8
:100260008610643083125820A60F1B29A1286300 32
:0202700037292C
```

```
:02400E00F53F7C
:00000001FF
```

The next project will be to develop a software routine to control the robot remotely. Remove the LCD display if it is still connected and plug the sonar ranger's power connector back into the +5V and ground connector on the controller board. This program is very similar to the last experiment where the light-emitting diodes were turned on and off depending on the position of the control stick. The only difference is that the robot will be given motion commands instead of light and sound commands. During the last experiment, you may have noticed that if the transmitter was turned off, the LEDs turned on and off in a random order. One of the conditions that needs to be considered when developing the remote control software for a moving robot is what kind of output the receiver produces when the radio transmitter is turned off. If we don't trap this situation at the beginning of the program, then the robot may act in an uncontrollable manner. Another situation that needs to be considered is what will happen if the transmitter drifts out of range, or is abruptly shut off while the robot is in motion. If the robot were being used on a reconnaissance mission gathering information on enemy troop movement and it dropped out of transmission range, the best thing to do would be for the robot to hide and then shut the systems down. The last thing you would want to happen would be for the robot to start running around erratically, drawing attention to itself, or wandering out into high-traffic areas. When it is determined that the remote control receiver is no longer receiving a signal, the robot's automatic processes could take over. The robot could then send out a distress beacon so that it could be located later. Another option would be for the robot to start executing mapping routines or find its way home using a GPS module and the obstacle-avoidance schemes that have been discussed.

When the transmitter control stick is in the centered home position, the output values are approximately 150. We can use this information to determine if the transmitter is powered up and within range. When the transmitter is turned off, the output values from the receiver are often zero with random numbers thrown in. When the transmitter is on, the output value from the receiver is never zero. When the program starts, it will make sure that the transmitter is turned on and the control stick is in the home position (the stick is brought back to home internally using a spring mechanism). If it is not, the robot will beep to notify the user that the transmitter needs to be turned on. Once the transmitter is turned on, the code will always check for zero values being produced by the receiver. If a zero value is picked up, then there is a problem with the transmitter or the receiver, and the program execution jumps back to warning the user that there is a problem by beeping. By comparing the values received to an acceptable window surrounding the value of 150, it can be determined if the transmitter is turned on as shown in the following code snippet:

```
trans_power:

pulsin chan_1,1,channel_1
pulsin chan_2,1,channel_2

 if (channel_1 > 155 or channel_1 < 145) or (channel_2 > 155 or
channel_2 < 145) then
  sound piezo,[100,5]
  goto trans_power
 endif
```

This code actually checks to see if the values are outside the acceptable window to make it easier to loop back and check again. When the values are within the proper range, the program continues on to the main section of code. When the main body of the code is being executed, the input values from the receiver are compared to zero before any other evaluations are done. If either of the values equals zero, then there is a problem with the transmitter or receiver, and the program execution jumps back up to the section of code to check for valid receiver input and notifies the user with a beeping noise.

Table 12.2 shows the values generated by the `pulsin` command for each radio channel, corresponding to the control stick when it is positioned at the maximum limit in each direction. Figure 12.8 shows the control stick positions and the robot movements that will be mapped to each position. Looking closely at Table 12.2 reveals that if the program is written in the right way, the only conditions that need to be evaluated are the walk forward, walk backward, turn left, and turn right, because all other movements are combinations of these four.

TABLE 12.2 Robot Movements Mapped to Remote Control Channel Values

Control Stick	Robot Action	Channel 1 Value	Channel 2 Value
Home	No motion	150	150
Forward	Walk forward	150	190
Reverse	Walk backward	150	105
Left	Turn left on the spot	190	150
Right	Turn right on the spot	105	150
Forward-right	Turn right walking forward	105	190
Reverse-right	Turn right walking reverse	105	105
Forward-left	Turn left walking forward	190	190
Reverse-left	Turn left walking backward	190	105

For example, if the control stick is moved into the position of forward-right (marked F-Right in Figure 12.8) then the value for channel 1 is 105 and the value for channel 2 is 190. To take care of this situation, the subroutine for walking forward will be called once, and the subroutine for turning right will be called once during one iteration or loop of program execution and radio input readings. By taking the receiver values at the beginning of the software routine and then mapping each of the four basic positional values to the correct movements, both of the walking subroutines needed for a forward right movement, or the other combination movements, will be executed. This is illustrated in the following code section:

```
pulsin chan_1,1,channel_1    ' get receiver values for channels 1 & 2
pulsin chan_2,1,channel_2

if channel_1 = 0 or channel_2 = 0 then    ' check to see if the
                                          ' transmitter is off
   goto trans_power                       ' or out of range.
endif

If channel_2 > 180 then                   ' signal to walk forward
   gosub walk_forward
endif

If channel_2 < 120 then                   ' signal to walk reverse
   gosub walk_reverse
endif

If channel_1 < 120 then                   ' signal to turn right
   gosub turn_right
endif

If channel_1 > 180 then                   ' signal to turn left
   gosub turn_left
endif
```

The entire remote-control.bas program is listed in Program 12.5. Compile the remote-control.bas program listed in Program 12.5 and program the PICmicro MCU 16F84 with the remote-control.hex file listed in Program 12.6. Power up the robot without turning on the transmitter. The robot should remain motionless, making beeping noises until the power to the transmitter is turned on. Turn on the power and try out the different directions listed in Figure 12.8. Notice that when you command the robot to move in a forward-right direction that the program is actually calling the subroutine to walk forward and then the subroutine to turn right. Switch off the power to the transmitter and see what happens. The robot should stop what it is doing and start beeping. The remote-control.bas program also uses

the sonar ranger to warn the user that the robot has come close to an object. If a wireless video camera with sound were added to the robot, this might be useful since it is difficult to judge depth in two dimensions from a video feed. The sonar could also be used to inhibit forward motion if an obstacle is encountered.

PROGRAM 12.5 **Remote-control.bas program listing.**

```
'*******************************************************************
' remote-control.bas
' Measure the width of the pulses produced by the radio receiver on
' channels 1 & 2.
' Use this information to determine the robots movements
' PicBasic Pro compiler - MicroEngineering Labs.
'*******************************************************************

trisa = %00000011          ' set porta pins 0&1 to inputs
trisb = %00001000          ' set portb pin 3 to input

m_servo      var   byte         ' define variables and constants
l_servo      var   byte
r_servo      var   byte
timer        var   byte
right_led    var   PORTB.0
left_led     var   PORTB.1
trigger      var   PORTB.2
echo         var   PORTB.3
piezo        var   PORTB.4
right_servo  var   PORTB.5
left_servo   var   PORTB.6
mid_servo    var   PORTB.7
chan_1       var   PORTA.0
chan_2       var   PORTA.1
Channel_1    var   byte
Channel_2    var   byte
dist_raw     var   word
dist_inch    var   word
conv_inch    con   15             ' conversion factor for inches
turn         var   byte
I            var   byte

low trigger                ' set sonar trigger low
low left_led               ' turn off left LED
low right_led              ' turn off right LED

sound piezo,[100,10,50,5,70,10,50,2]     ' Make startup sound
```

```
trans_power:

      pulsin chan_1,1,channel_1      ' check to see if the transmitter
      pulsin chan_2,1,channel_2      ' is turned on or not

      if (channel_1 > 155 or channel_1 < 145) or (channel_2 > 155
        or channel_2 < 145) then
          sound piezo,[100,5]
          goto trans_power          ' if transmitter is off then keep
                                    ' checking
      endif

main:

      pulsin chan_1,1,channel_1     ' get receiver values for channels
                                    ' 1 & 2
      pulsin chan_2,1,channel_2

      if channel_1 = 0 or channel_2 = 0 then    ' check to see if the
                                                ' transmitter is off
          goto trans_power                      ' or out of range.
      endif

      If channel_2 > 180 then                  ' signal to walk forward
          gosub walk_forward
      endif

      If channel_2 < 120 then                  ' signal to walk reverse
          gosub walk_reverse
      endif

      If channel_1 < 120 then                  ' signal to turn right
          gosub turn_right
      endif

      If channel_1 > 180 then                  ' signal to turn left
          gosub turn_left
      endif

      gosub sr_sonar                ' get sonar ranger information
      if dist_inch < 6 then         ' check to see if the robot is
                                    ' within 5 inches of
          high left_led             ' an object. Make warning noise and
                                    ' flash LEDs if
          high right_led            ' an obstacle is encountered.
          sound piezo,[80,10,100,5,110,2,90,2]
```

```
        low left_led
        low right_led
    endif

Goto main

walk_forward:                               ' walk forward subroutine
     m_servo = 170
     gosub servo
     l_servo = 160
     r_servo = 160
     gosub servo
     m_servo = 100
     gosub servo
     l_servo = 120
     r_servo = 120
     gosub servo
return

walk_reverse:                               ' walk reverse subroutine
     m_servo = 170
     gosub servo
     l_servo = 120
     r_servo = 120
     gosub servo
     m_servo = 100
     gosub servo
     l_servo = 160
     r_servo = 160
     gosub servo
return

turn_left:                          ' rotate the robot left 18 degrees
                  m_servo = 170
                  gosub servo
                  l_servo = 120
                  r_servo = 160
                  gosub servo
                  m_servo = 100
                  gosub servo
                  l_servo = 160
                  r_servo = 120
                  gosub servo
return

turn_right:                         ' rotate the robot right 18 degrees
                  m_servo = 170
```

```
                  gosub servo
                  l_servo = 160
                  r_servo = 120
                  gosub servo
                  m_servo = 100
                  gosub servo
                  l_servo = 120
                  r_servo = 160
                  gosub servo
return

sr_sonar:

         pulsout trigger,1                       ' send a 10us trigger
                                                 ' pulse to the SRF04
         pulsin echo,1,dist_raw                  ' start timing the
                                                 ' pulse width on echo
                                                 ' pin
         dist_inch = (dist_raw/conv_inch)        ' Convert raw data
                                                 ' into inches
         pause 1                                 ' wait for 10us before
                                                 ' returning to main

return

servo:                                       ' subroutine to set servos
       for timer = 1 to 10
       pulsout mid_servo,m_servo
       pulsout left_servo,l_servo
       pulsout right_servo,r_servo
       pause 13
       next timer
return

End
```

PROGRAM 12.6 Remote-control.hex file listing.

```
:10000000DB28A8008417800484138E010C1C8E004A
:1000100023200319D62823200319D6282320D628E5
:10002000A80059200C080D040319D628D0208413E9
:100030002808800664001C281D288C0A03198D0FCF
:100040001A288006D62828088E0601308C008D01DB
:1000500000828050E06031D08008C0A03198D0FE1
:100060002828080008F002A08840028095A208413B1
:100070008F080319D628F03091000E0880389000C0
```

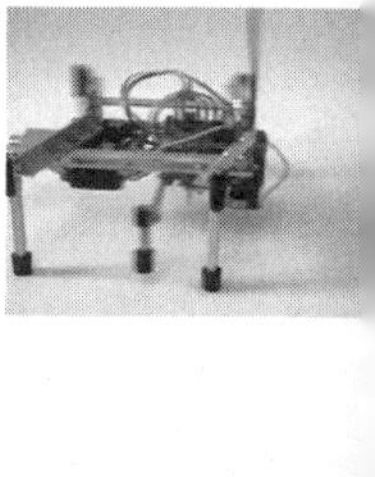

```
:10008000F03091030319910003198F030319D62847
:1000900049285D2003010C1828088E1F28088E08A7
:1000A00003190301900F562880063D2857280000A9
:1000B0004028FF3A84178005D6280D080C04031940
:1000C0008C0A80300C1A8D060C198D068C188D0642
:1000D0000D0D8C0D8D0DD6288F018E00FF308E07F3
:1000E00031C8F07031CD62803308D00DF307A20D5
:1000F0006E288D01E83E8C008D09FC30031C83289E
:100100008C07031880288C0764008D0F80280C183A
:1001100089288C1C8D2800008D2808008D018F01F6
:100120008E000230A0288E000330A0288D018F01A0
:100130008E000130A0288D018F018E000430A02890
:1001400094000F080D02031DA7280E080C020430AE
:100150000318013003190230140503 1DFF30D6289F
:10016000040 4031DFF30D628910190011030920045
:100170000D0D900D910D0E0890020F08031C0F0F2E
:1001800091020318CA280E0890070F0803180F0FD2
:1001900091070310 8C0D8D0D920BB8280C08D628F2
:1001A0008C098D098C0A03198D0A08008313031327
:1001B00083126400080083160330850008308600 2F
:1001C0008312061183160611831286108316861079
:1001D0008312061083160610063 08312AA00103010
:1001E000A80064308E000A30322032308E00053094
:1001F00032204630 8E000A30322032308E000230FB
:100200003220013 08C000530840001300120 0C08C0
:10021000B00001308C00053084000230012 00C0851
:10022000B10030088C009B3096209E0030088C0076
:1002300091309B20A0001E0884002008B020A00060
:10024000A10031088C009B309620A20031088C0060
:1002500091309B20A40022088400240 8B020A40030
:10026000A50020082104840024082504B020A4004F
:10027000A50064002408250403194729063 0AA00B4
:100280001030A80064308E00053032200129013082
:100290008C000530840001300120 0C08B0000130D2
:1002A0008C0005308400023001200C08B1003008B9
:1002B0008C0000308E209E0031088C0000308E2093
:1002C000A0001E0884002008B020A000A100640047
:1002D00020082104031 96D2901296400B530310279
:1002E00031C7329B421640078303102031 8792982
:1002F000C5216400783030020318 7F29E7216400AB
:10030000B5303002031C8529D621F8212C088C0039
:100310002D088D008F0106309320031DB32986140C
:10032000831686108312061483160610063083127 5
:10033000AA001030A80050308E000A3032206430FD
:100340008E00053032206E308E00023032205A305E
:100350008E0002303220861083168610831206101B
:100360008316061083124729AA30B4001722A03042
```

```
:10037000B300A030B50017226430B40017227830E3
:10038000B3007830B50017220800AA30B400172255
:100390007830B3007830B50017226430B4001722EB
:1003A000A030B300A030B50017220800AA30B40076
:1003B00017227830B300A030B50017226430B400A3
:1003C0001722A030B3007830B50017220800AA30F9
:1003D000B4001722A030B3007830B5001722643083
:1003E000B40017227830B300A030B50017220800FF
:1003F00001308C008D01063084000430102001306 3
:100400008C0006308400083001200C08AE000D0876
:10041000AF002E088C002F088D000F308E008F014A
:10042000B420AC000D08AD0001306C200800013094
:10043000B60064000B3036020318372A34088C00EB
:100440008D0106308400803010203 3088C008D012F
:10045000006308400403010203 5088C008D010630B5
:10046000840020301020 0D306C20B60F192A0800AF
:040470006300382AC3
:02400E00F53F7C
:00000001FF
```

While running the `remote-control.bas` program to remotely pilot the robot, you may have noticed that the robot's movements are a little rough when the control stick is in a position that requires two different subroutines to be called. This can be improved by writing a motion subroutine for each movement command. In the `remote-control.bas` program, when the control stick is in the forward-right position, the subroutine to walk forward is called, then the subroutine to turn right is called. This could be combined into one subroutine that flows smoothly. Instead of briefly stopping the robot to turn right while walking forward, the subroutine could extend the leg movement on the robot's left side and inhibit the leg movement a little on the right. Using this scheme, the robot would continue to move forward, but would also move to the right at the same time. All of the different conditions listed in Table 12.2 will need to be evaluated and then the appropriate walking subroutine called. The next program is called `conditions-remote.bas` and is listed in Program 12.7. Program the PIC 16F84 with the corresponding `conditions-remote.hex` file listed in Program 12.8. You will notice that the robot now walks smoothly when commanded to move in one of the combined directions, because each now has its own subroutine.

PROGRAM 12.7   Conditions-remote.bas program listing.

```
'****************************************************************
' conditons-remote.bas
' Robot uses seperate walking subroutines for each of the 8
```

```
' direction commands.
' The robot walks in a smooth, coordinated manner.
' PicBasic Pro compiler - MicroEngineering Labs.
'**************************************************************

trisa = %00000011                ' set porta pins 0&1 to inputs
trisb = %00001000                ' set portb pin 3 to input

m_servo      var   byte          ' define variables and constants
l_servo      var   byte
r_servo      var   byte
timer        var   byte
right_led    var   PORTB.0
left_led     var   PORTB.1
trigger      var   PORTB.2
echo         var   PORTB.3
piezo        var   PORTB.4
right_servo  var   PORTB.5
left_servo   var   PORTB.6
mid_servo    var   PORTB.7
chan_1       var   PORTA.0
chan_2       var   PORTA.1
Channel_1    var   byte
Channel_2    var   byte
dist_raw     var   word
dist_inch    var   word
conv_inch    con   15              ' conversion factor for inches
turn         var   byte
i            var   byte

low trigger
  ' set sonar trigger low
low left_led                     ' turn off left LED
low right_led                    ' turn off right LED

sound piezo,[100,10,50,5,70,10,50,2]     ' Make startup sound

trans_power:

     pulsin chan_1,1,channel_1    ' check to see if the transmitter
     pulsin chan_2,1,channel_2    ' is turned on or not

     if (channel_1 > 155 or channel_1 < 145) or (channel_2 > 155
       or channel_2 < 145) then
        sound piezo,[100,5]
        goto trans_power          ' if transmitter is off then keep
                                  ' checking
```

```
      endif

main:

      pulsin chan_1,1,channel_1                  ' get receiver values
                                                 ' for channels 1 & 2
      pulsin chan_2,1,channel_2

      if channel_1 = 0 or channel_2 = 0 then     ' check to see if the
                                                 ' transmitter is off
         goto trans_power                        ' or out of range.
      endif

      If channel_2 > 180 and (channel_1 > 145 and channel_1 < 155)
       then
         gosub walk_forward
      endif

      If channel_2 < 120 and (channel_1 > 145 and channel_1 < 155)
       then
         gosub walk_reverse
      endif

      If channel_1 < 120 and (channel_2 > 145 and channel_2 < 155)
       then
         gosub turn_right
      endif

      If channel_1 > 180 and (channel_2 > 145 and channel_2 < 155)
       then
         gosub turn_left
      endif

      If channel_1 < 120 and channel_2 > 180 then
         gosub right_forward
      endif

      If channel_1 > 180 and channel_2 > 180 then
         gosub left_forward
      endif

     If channel_1 < 120 and channel_2 < 120 then
         gosub right_reverse
      endif

     If channel_1 > 180 and channel_2 < 120 then
         gosub left_reverse
```

```
      endif

      gosub sr_sonar              ' get sonar ranger information
      if dist_inch < 6 then       ' check to see if the robot is
                                  ' within 5 inches of
         high left_led            ' an object. Make warning noise and
                                  ' flash LEDs if
         high right_led           ' an obstacle is encountered.
         sound piezo,[80,10,100,5,110,2,90,2]
         low left_led
         low right_led
      endif

Goto main

walk_forward:                     ' walk forward subroutine
     m_servo = 170
     gosub servo
     l_servo = 160
     r_servo = 160
     gosub servo
     m_servo = 100
     gosub servo
     l_servo = 120
     r_servo = 120
     gosub servo
return

walk_reverse:                     ' walk reverse subroutine
     m_servo = 170
     gosub servo
     l_servo = 120
     r_servo = 120
     gosub servo
     m_servo = 100
     gosub servo
     l_servo = 160
     r_servo = 160
     gosub servo
return

turn_left:                  ' rotate the robot left 18 degree
                  m_servo = 170
                  gosub servo
                  l_servo = 120
                  r_servo = 160
                  gosub servo
```

```
                    m_servo = 100
                    gosub servo
                    l_servo = 160
                    r_servo = 120
                    gosub servo
return

turn_right:                          ' rotate the robot right 18 degrees
                    m_servo = 170
                    gosub servo
                    l_servo = 160
                    r_servo = 120
                    gosub servo
                    m_servo = 100
                    gosub servo
                    l_servo = 120
                    r_servo = 160
                    gosub servo
return

right_forward:                                ' forward right subroutine
     m_servo = 170
     gosub servo
     l_servo = 160
     r_servo = 140
     gosub servo
     m_servo = 100
     gosub servo
     l_servo = 120
     r_servo = 130
     gosub servo
return

left_forward:                                  ' forward left subroutine
     m_servo = 170
     gosub servo
     l_servo = 140
     r_servo = 160
     gosub servo
     m_servo = 100
     gosub servo
     l_servo = 130
     r_servo = 120
     gosub servo
return

right_reverse:                                 ' right reverse subroutine
```

```
    m_servo = 170
    gosub servo
    l_servo = 130
    r_servo = 120
    gosub servo
    m_servo = 100
    gosub servo
    l_servo = 140
    r_servo = 160
    gosub servo
return

left_reverse:                                   ' left reverse subroutine
    m_servo = 170
    gosub servo
    l_servo = 120
    r_servo = 130
    gosub servo
    m_servo = 100
    gosub servo
    l_servo = 160
    r_servo = 140
    gosub servo
return

sr_sonar:

         pulsout trigger,1                        ' send a 10us trigger
                                                  ' pulse to the SRF04
         pulsin echo,1,dist_raw                   ' start timing the
                                                  ' pulse width on echo
                                                  ' pin
         dist_inch = (dist_raw/conv_inch)         ' Convert raw data
                                                  ' into inches
         pause 1                                  ' wait for 10us before
                                                  ' returning to main

return

servo:                                          ' subroutine to set servos
       for timer = 1 to 10
       pulsout mid_servo,m_servo
       pulsout left_servo,l_servo
       pulsout right_servo,r_servo
       pause 13
       next timer
return

End
```

PROGRAM 12.8 Conditions-remote.hex file listing.

```
:10000000E228A8008417800484138E010C1C8E0043
:1000100023200319DD2823200319DD282320DD28D0
:10002000A80059200C080D040319DD28D7208413DB
:100030002808800664001C281D288C0A03198D0FCF
:100040001A288006DD2828088E0601308C008D01D4
:10005000000828050E06031D08008C0A03198D0FE1
:1000600028280800 8F002A08840028095A208413B1
:100070008F080319DD28F03091000E0880389000B9
:10008000F03091030319910003198F030319DD2840
:1000900049285D2003010C1828088E1F28088E08A7
:1000A00003190301900F562880063D2857280000A9
:1000B0004028FF3A84178005DD280D080C04031939
:1000C0008C0A80300C1A8D060C198D068C188D0642
:1000D0000D0D8C0D8D0DDD288F018E00FF308E07EC
:1000E000031C8F07031CDD2803308D00DF307A20CE
:1000F0006E288D01E83E8C008D09FC30031C83289E
:100100008C07031880288C0764008D0F80280C183A
:1001100089288C1C8D2800008D2808008D018F01F6
:100120008E000230A0288E000330A0288D018F01A0
:100130008E000130A0288D018F018E000430A02890
:1001400094000F080D02031DA7280E080C020430AE
:10015000031801300319023014 05031DFF30DD2898
:10016000000380 31DFF300405031DFF30DD280404A3
:10017000031DFF30DD2891019001103092000D0D1C
:100180000900D910D0E0890020F08031C0F0F9102A5
:100190000000318D1280E0890070F0803180F0F9107B6
:1001A00003108C0D8D0D920BBF280C08DD288C09D7
:1001B0008D098C0A03198D0A08008313031383121 7
:1001C00006400080 08316033085000830860083121F
:1001D00006118316061183128610831686108312 69
:1001E000061083160610063083 12AA001030A800ED
:1001F00064308E000A30322032308E0005303220DA
:1002000046308E000A30322032308E0002303220EA
:1002100001308C0005308400013001200C08B00052
:1002200001308C0005308400023001200C08B10040
:1002300030088C009B3096209E0030088C00913056
:100240009B20A0001E0884002008B720A000A10069
:1002500031088C009B309620A20031088C00913030
:100260009B20A40220884002408B720A400A50035
:10027000020082104840024082504B720A400A50038
:10028000640024082504031 94E290630AA00103002
:10029000A80064308E0005303220082901308C001F
:1002A0000530840001300 1200C08B00001308C00C2
:1002B0000530840002300 1200C08B10030088C00A9
:1002C00000 0308E209E0031088C0000308E20A0006F
```

```
:1002D0001E0884002008B720A000A10064002008A8
:1002E00021040319742908293 1088C00B4309620A0
:1002F0009E0030088C0091309620A00030088C00C1
:100300009B309B20A200200884002208B020A2007D
:10031000A3001E0884002208 2304B020A200A3002A
:1003200064002208230403199629832231088C00D3
:1003300078309B209E0030088C0091309620A000E1
:1003400030088C009B309B20A200200884002208EB
:10035000B020A200A3001E08840022082304B020BD
:10036000A200A300640022082304 0319B8299422E0
:1003700030088C0078309B209E0031088C00913032
:100380009620A00031088C009B309B20A200200802
:1003900084002208B020A200A3001E0884002208C6
:1003A0002304B020A200A3006400220823040319 40
:1003B000DA29B62230088C00B43096209E0031082D
:1003C0008C0091309620A00031088C009B309B203F
:1003D000A200200884002208B020A200A3001E086A
:1003E000840022082304B020A200A3006400220895
:1003F00023040319FC29A52230088C0078309B20A7
:100400009E0031088C00B4309620A0001E088400A5
:100410002008B020A000A1006400200821040319D6
:10042000122AC72230088C00B43096209E00310872
:100430008C00B4309620A0001E0884002008B02054
:10044000A000A1006400200821040319282AD82252
:1004500030088C0078309B209E0031088C0078306A
:100460009B20A0001E0884002008B020A000A1004E
:1004700064002008210403193E2AE92230088C0078
:10048000B43096209E0031088C0078309B20A0006C
:100490001E0884002008B020A000A10064002008ED
:1004A00021040319542AFA220B232C088C002D084E
:1004B0008D008F0106309320031D822A8614831637
:1004C000861083120614831606100630 8312AA00C3
:1004D0001030A80050308E000A30322064308E0078
:1004E000053032206E308E00023032205A308E00BD
:1004F00002303220861083168610831206108316 6F
:1005000006108312 4E29AA30B4002A23A030B3006B
:10051000A030B5002A236430B4002A237830B30019
:100520007830B5002A230800AA30B4002A23783096
:10053000B3007830B5002A236430B4002A23A030F9
:10054000B300A030B5002A230800AA30B4002A2343
:100550007830B300A030B5002A236430B4002A23D9
:10056000A030B3007830B5002A230800AA30B400C8
:100570002A23A030B3007830B5002A236430B400B9
:100580002A237830B300A030B5002A230800AA300F
:10059000B4002A23A030B3008C30B5002A23643085
:1005A000B4002A237830B3008230B5002A23080033
:1005B000AA30B4002A238C30B300A030B5002A231F
```

```
:1005C0006430B4002A238230B3007830B5002A2387
:1005D0000800AA30B4002A238230B3007830B50076
:1005E0002A236430B4002A238C30B300A030B50035
:1005F0002A230800AA30B4002A237830B3008230BE
:10060000B5002A236430B4002A23A030B3008C3014
:10061000B5002A23080001308C008D0106308400CB
:100620000430102001308C00063084000830012096
:100630000C08AE000D08AF002E088C002F088D00AE
:100640000F308E008F01BB20AC000D08AD000130D3
:100650006C2008000130B60064000B30360203182D
:100660004A2B34088C008D01063084008030102025
:1006700033088C008D0106308400403010203508 8E
:100680008C008D0106308400203010200D306C204D
:0A069000B60F2C2B080063004B2B63
:02400E00F53F7C
:00000001FF
```

Summary

In this chapter, a remote control has been interfaced to the robot to control its movements from a distance of up to one mile. If a wireless video camera were added, the operator would not even have to keep the robot in sight to control it. In Chapter 13, we will take advantage of the extra radio channels and add a robotic gripper so that the robot will be able to pick up objects and move them around under remote control.

13

ADDING A ROBOTIC GRIPPER

Now that the robot can be remotely controlled, a gripper will be added so that objects can be picked up and moved from one location to another. Figure 13.1 is a picture of the lightweight gripper that will be constructed using aluminum and a standard servo. I designed this gripper to use the parallel linkage concept to keep the gripper fingers parallel throughout the horizontal motion. The unique feature of this design is that the rotary motion of the servo is converted to a more linear motion by using mechanical linkages so that gears do not need to be used. The gripper will be controlled directly from the radio receiver on one of the unused radio channels.

The great thing about already having a four-channel radio transmitter and receiver with the robot is that there are still two channels that can be used. The channel that is normally used as the throttle control for an airplane will be used to directly control the gripper. Unlike the control stick on the right side of the transmitter, the throttle control remains in the last position where it was placed, so that the airplane speed remains constant. This works out well for a robot gripper because once the operator has the gripper applying the right amount of pressure to an object, he can concentrate on controlling the robot's direction. For this project, the SRF04 sonar ranger will need to be removed before the gripper can be added. The parts needed to fabricate the gripper are listed in Table 13.1.

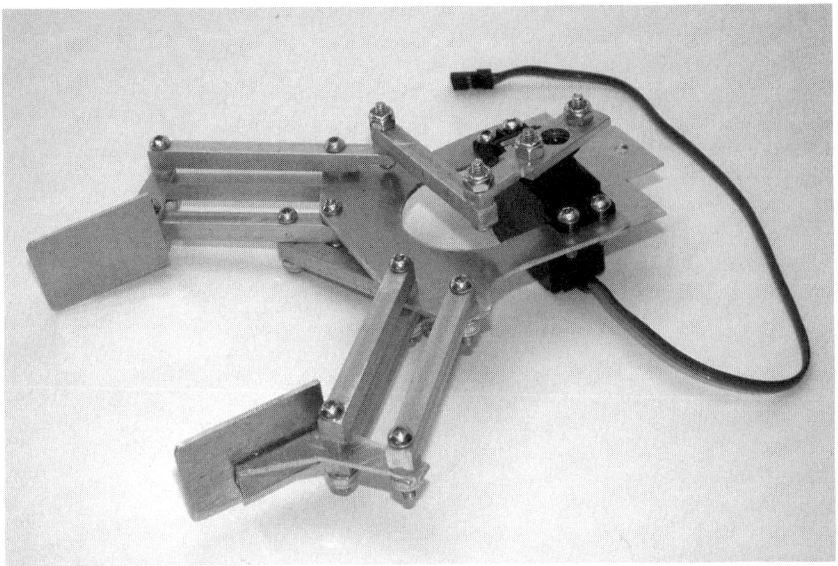

FIGURE 13.1 Robotic gripper that will be added to the robot.

TABLE 13.1 Gripper Parts List

PART	QUANTITY	DESCRIPTION
Standard servo	1	Servo, horn and screw
1/16-inch thick aluminum	10 inches × 3 inches	Aluminum stock
1/4 inch x 1/4-inch aluminum	25 inches	Aluminum stock
1/2 inch x 1/8-inch aluminum	2 inches	Aluminum stock
1-inch length machine screw	4	6/32 diameter
3/4-inch length machine screw	7	6/32 diameter
1/2-inch length machine screw	6	6/32 diameter
6/32 nylon washer	14	Nylon washer
6/32 diameter lock nut	17	Locking nut

The first part of the gripper that needs to be constructed is the base section. The standard servo and all other parts will be mounted to the gripper base. Cut a piece of 1/16-inch thick aluminum to a size of 3 1/2 × 2 3/4 inches, as shown in Figure 13.2. Photocopy the image in Figure 13.2 onto a sheet of paper and use the enlarge feature until the dotted outline is exactly 3 1/2 × 2 3/4 inches. Alternatively, you can scan the image into your computer and use a graphics editor program to make the enlargement and then print the image or download the file from my Web site (http://www.thinkbotics.com/). Cut the image out and glue it to the 3 1/2 × 2 3/4-inch piece of aluminum. Cut the aluminum along the guide lines with a hacksaw or metal-cutting band saw. Drill the holes where marked with a 5/32-inch drill bit. Peel the paper off the aluminum and file the edges smooth. The finished part is shown in Figure 13.3, and is marked as piece A (Figure 13.2). Use two 1/2-inch, 6/32 diameter machine screws and lock nuts to secure the standard servo in place at the two mounting holes closest to the front of the gripper, as shown in Figure 13.3.

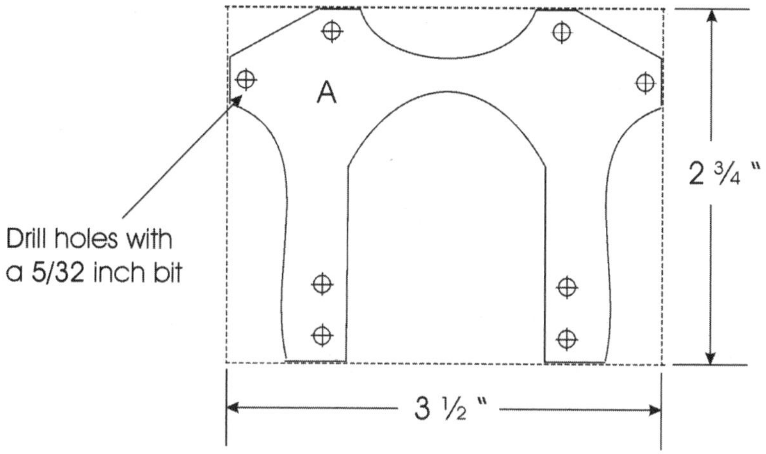

FIGURE 13.2 Cutting and drill guide for the gripper base.

FIGURE 13.3 Finished gripper base with servo installed.

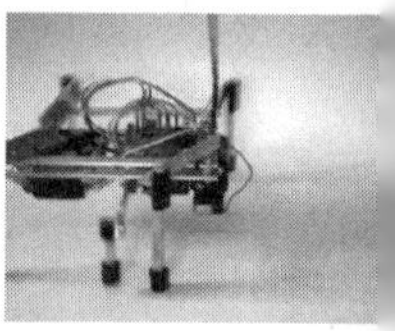

Next, the gripper fingers will be fabricated. Cut two pieces of 1/16-inch thick aluminum each 1 5/8 inches wide × 1 5/8 inches long. Photocopy the cutting and drilling overlay shown in Figure 13.4, and enlarge until the dimensions are exactly 1 5/8 × 1 5/8 inches. Cut the templates out and glue them onto the aluminum. Use a hacksaw to cut the pieces out along the guide lines. Drill the holes as marked with a 5/32-inch drill bit. Figure 13.4 shows where the aluminum is bent on 90-degree angles downward. Note that part C is a mirror copy of B. Place each piece in a table vise, and bend by tapping lightly with a hammer. See Figure 13.5, which shows the finished finger pieces as a guide for bending. Peel the paper template from the aluminum, file the edges smooth, and mark the pieces with the letters B and C.

Using the 1/16-inch thick aluminum, cut out pieces D and E as shown in Figure 13.4. File the edges smooth and then mark the parts with the corresponding letters.

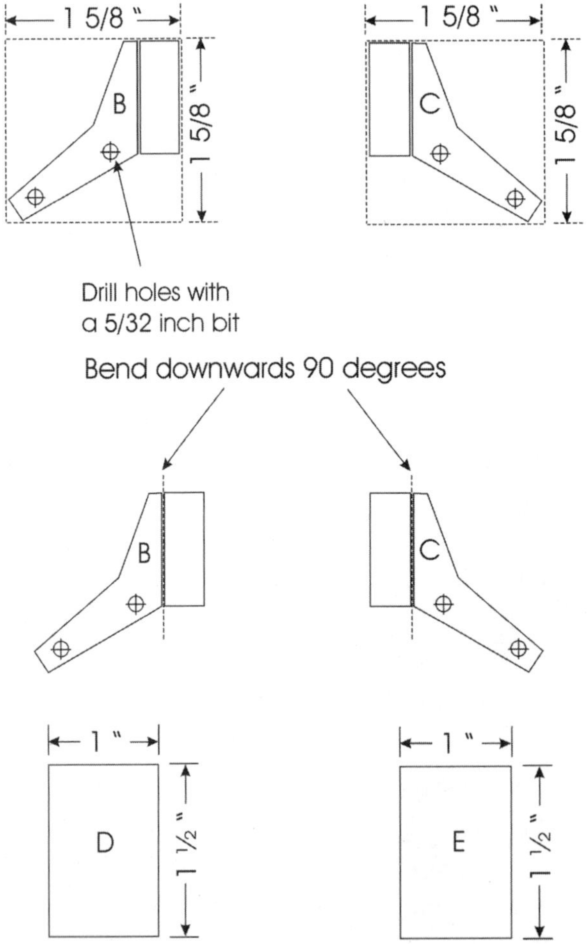

FIGURE 13.4 **Cutting and drilling guide for gripper finger pieces.**

The part labeled B will now be glued to part D, and part C will be glued to part E. Lay part D flat on a table, and use a hot glue gun to mount part B in the lower left corner of part D. See Figure 13.5 for an idea of where the part is mounted. Do the same with parts C and E, except that part C is glued into the bottom right side of part E. See Figure 13.5 for placement and gluing details.

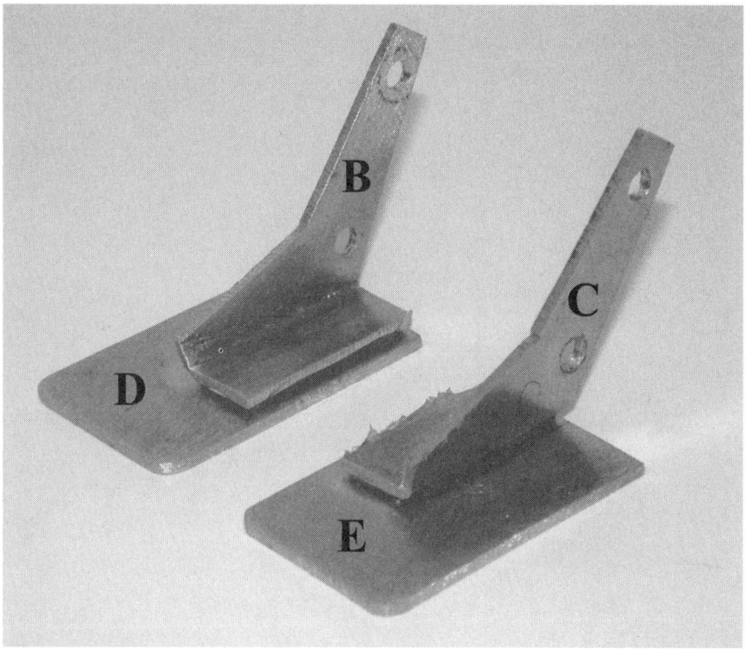

FIGURE 13.5 Finished gripper fingers.

Cut four pieces of the 1/4-inch × 1/4-inch aluminum 2 1/2 inches long each, as shown in Figure 13.6. Three of the pieces are marked as F, G, and H, and only have two holes drilled in each with a 5/32-inch bit. The fourth piece is labeled I, and has an extra hole also drilled with a 5/32-inch bit as shown in the diagram.

Cut two pieces from the 1/4-inch × 1/4-inch aluminum stock to a length of 2 1/4 inches, and then drill the holes as indicated in Figure 13.6 with a 5/32-inch bit. These two pieces are labeled J and K. Figure 13.7 shows the finished cut and drilled pieces.

Fabricate two pieces marked as L and M cut from the 1/4-inch × 1/4-inch aluminum stock to a length of 3 inches each. See Figure 13.8 for the cutting guide and drilling positions. Once the pieces have been cut and drilled, the end with an extra hole located at 5/8 inch will need to be modified. Take the piece marked as L, measure 1/2 inch from the end of the piece, and mark with a pencil. Use a hacksaw to cut into the aluminum, along the mark made with the pencil, to a depth of 1/8 inch. Turn

the piece on its end and measure and mark 1/8 inch from the side with the hole. Hold the piece in a vise and cut from the end until the piece is removed. Do the same for the second piece, marked M. The finished pieces should look like the side view illustration in Figure 13.8, and the photograph in Figure 13.9.

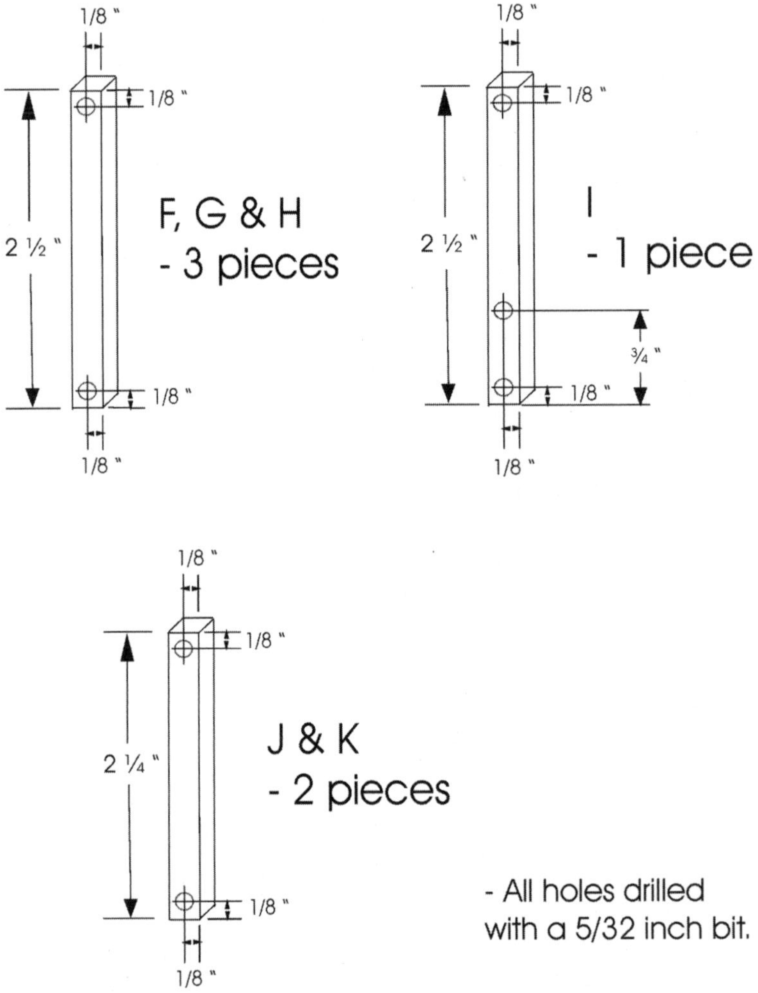

FIGURE 13.6 Gripper linkages cutting and drilling guide.

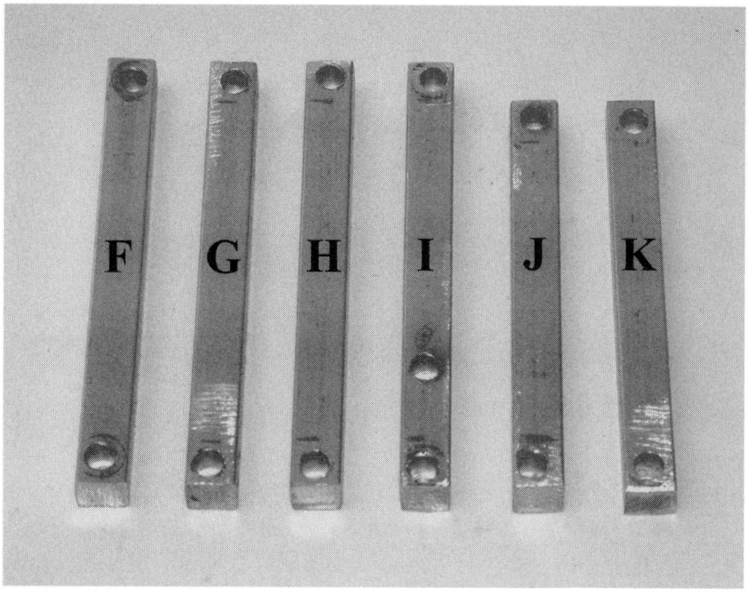

FIGURE 13.7 Finished gripper linkage pieces.

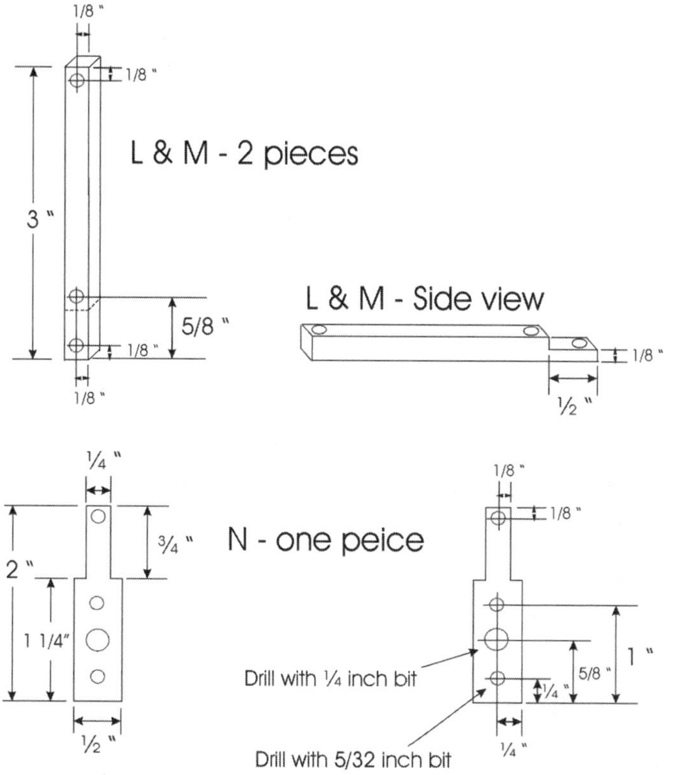

FIGURE 13.8 Gripper linkages and servo mount cutting and drilling guide.

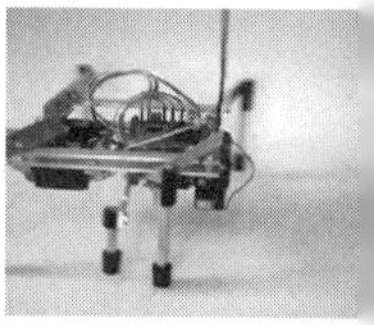

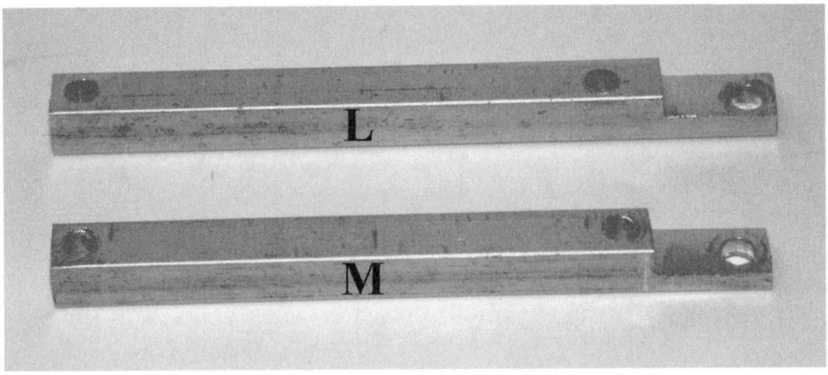

FIGURE 13.9 Finished linkage pieces L and M.

The next step is to build the servo horn mount, marked as piece N. Use a piece of 1/2-inch × 1/8-inch aluminum stock cut to a length of 2 inches. Figure 13.8 shows the cutting dimensions and the hole drilling measurements. Modify a servo horn by drilling 5/32-inch mounting holes that align with the holes in piece N when the center holes of the servo horn and piece N are lined up. Cut off any excess plastic so that the servo horn is flush with the aluminum piece. Figure 13.10 shows the finished aluminum servo horn mount along with the modified servo horn.

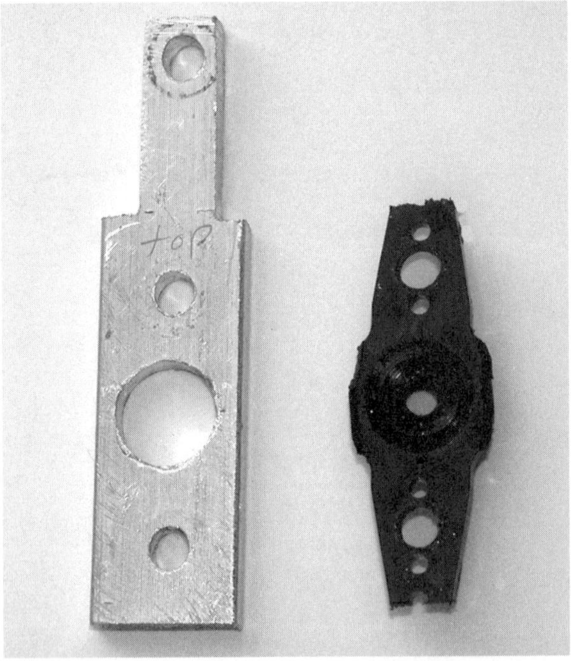

FIGURE 13.10 Servo mount and modified servo horn.

Mount the servo horn to piece N using two 1/2-inch machine screws and lock nuts as shown in Figure 13.11.

FIGURE 13.11 Finished mount attached to servo horn.

Assembling the Parts

Figure 13.12 provides an overview of how the parts are assembled to make up the gripper. Each step will be covered in detail in the following section.

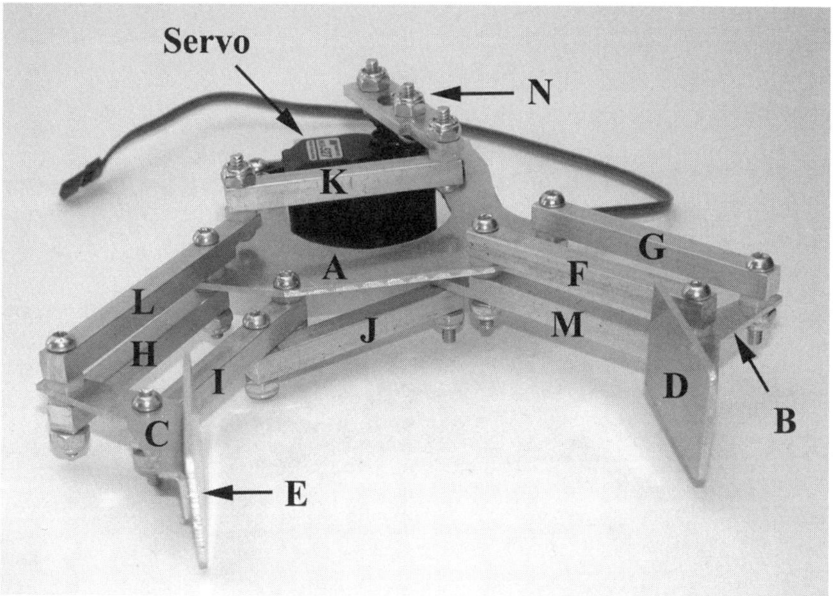

FIGURE 13.12 Individual parts assembled to make up the robotic gripper.

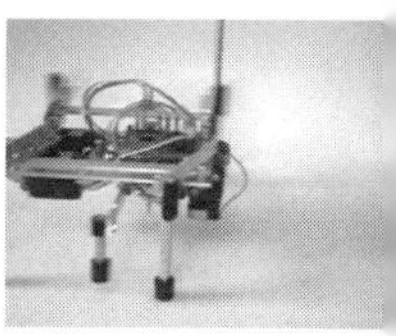

Start by attaching parts L and K, as shown in Figure 13.13. Place a 6/32-inch nylon washer between the two pieces, and secure with a 3/4-inch machine screw and lock nut. Tighten the lock nut with just enough pressure so that the pieces move freely without any effort, but do not flop around.

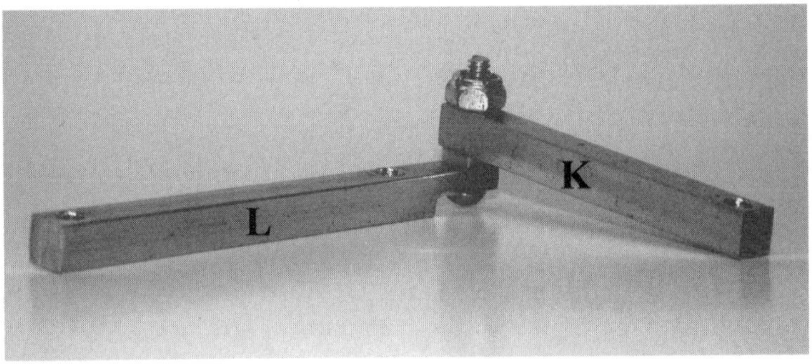

FIGURE 13.13 **Subassembly made up of parts L and K.**

Attach the subassembly (made up of pieces L and K) to the gripper base along with part H using a 1-inch, 6/32 diameter machine screw and lock nut. Place a nylon washer between piece L and the gripper base. Attach piece K to N with a 3/4-inch machine screw and lock nut, as shown in Figure 13.14. Place a nylon washer between part K and N. Part K goes underneath part N.

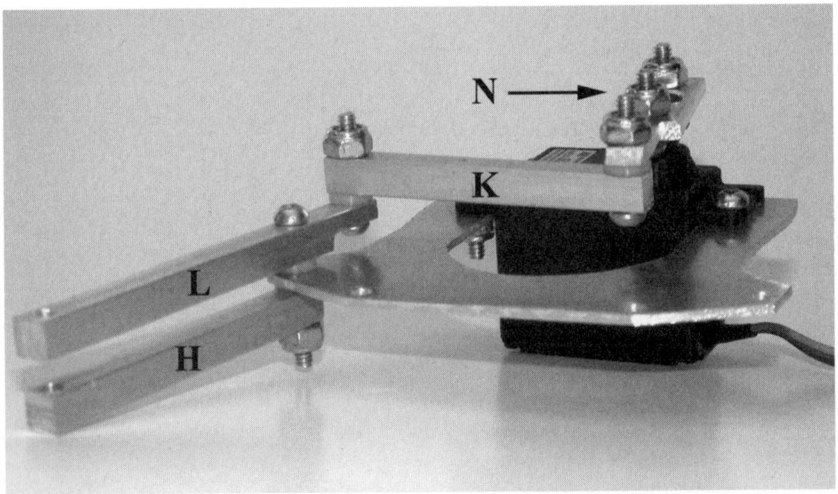

FIGURE 13.14 **Parts L, K, and H attached to gripper base and servo linkage N.**

The next subassembly is made up of parts I, J, and M. Use Figure 13.15 as a guide to assemble the parts. Place a nylon washer between parts M and J and I and J. Use two 3/4-inch machine screws and lock nuts to assemble. Again, tighten with enough pressure to let the parts move freely.

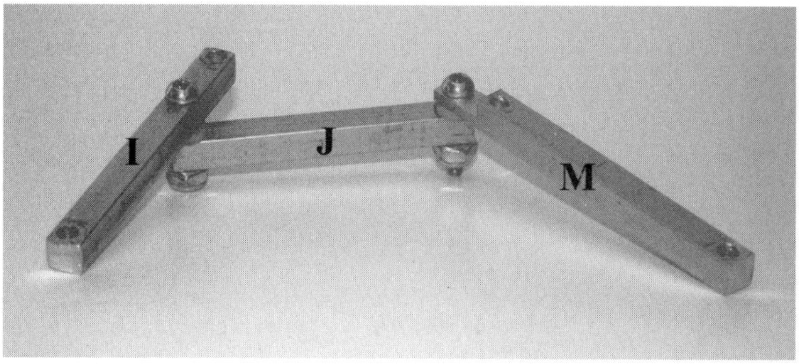

FIGURE 13.15 **Subassembly made up of parts I, J, and M.**

Attach the subassembly made up of parts I, J, and M along with part F to the robot base with a 3/4-inch machine screw for part I, and a 1-inch machine screw for part M and F. Place nylon washers between parts I, M, and F and the gripper base. Secure with lock nuts but let the parts move freely. Figure 13.16 shows how the parts should be assembled.

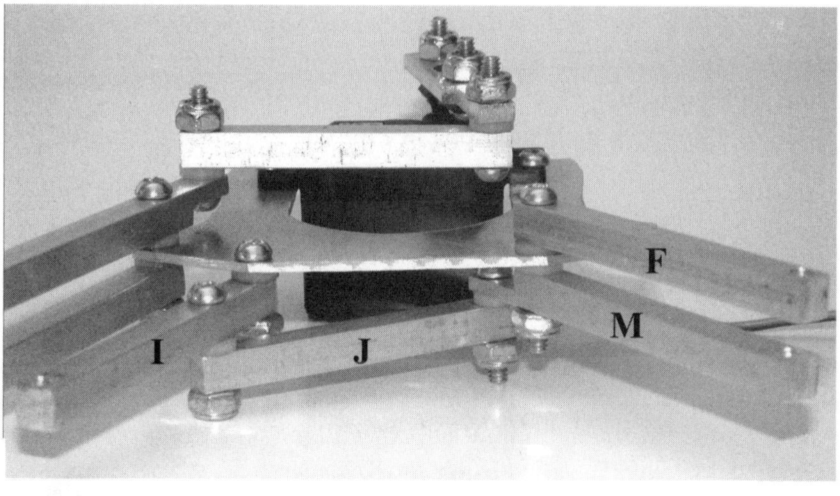

FIGURE 13.16 **Parts I, J, M, and F fastened to the gripper base.**

Attach the gripper finger piece made up of parts C and E to linkage pieces H, I, and L using Figure 13.17 as a guide. Use nylon washers to separate the pieces. Secure the finger to piece I with a 3/4-inch, 6/32 diameter machine screw and lock nut. Use a 1-inch, 6/32 diameter machine screw and locking nut for pieces L and H. Tighten the lock nuts with enough pressure so that the finger pieces move freely.

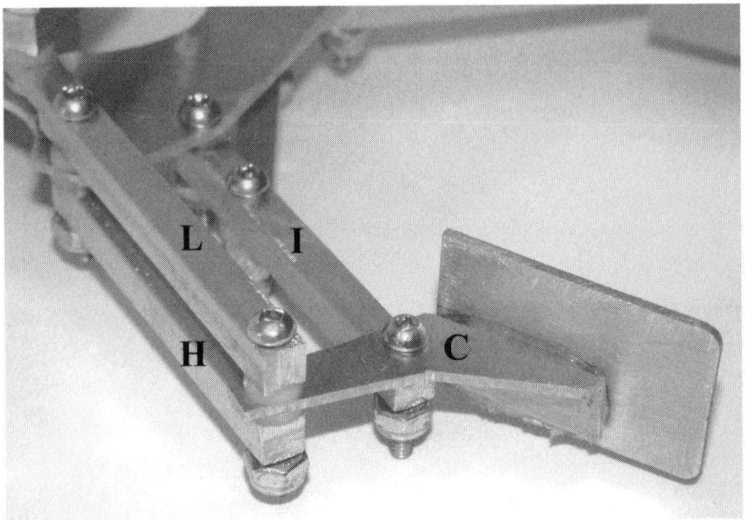

FIGURE 13.17 **Gripper finger C attached to mechanical linkage pieces H, I, and L.**

Attach the gripper finger piece made up of parts B and D to linkage pieces F, G, and M using Figure 13.18 as a guide. Use nylon washers to separate the pieces. Secure the finger to piece G with a 3/4-inch, 6/32 diameter machine screw and lock nut. Use a 1-inch, 6/32 diameter machine screw and lock nut for pieces F and M.

The fully assembled gripper should now look like the one pictured in Figure 13.12.

Mounting the Gripper to the Robot

In order to mount the gripper to the robot, a mounting bracket will be fabricated using a piece of the 1/16-inch thick aluminum. Follow the cutting and drilling guide in Figure 13.19 to construct the mounting bracket. Attach the mounting bracket to the gripper base where the servo is attached using two 1/2-inch, 6/32 diameter machine screws and lock washers as shown in Figure 13.20. The machine screws used to attach the mounting bracket are shared by the servo.

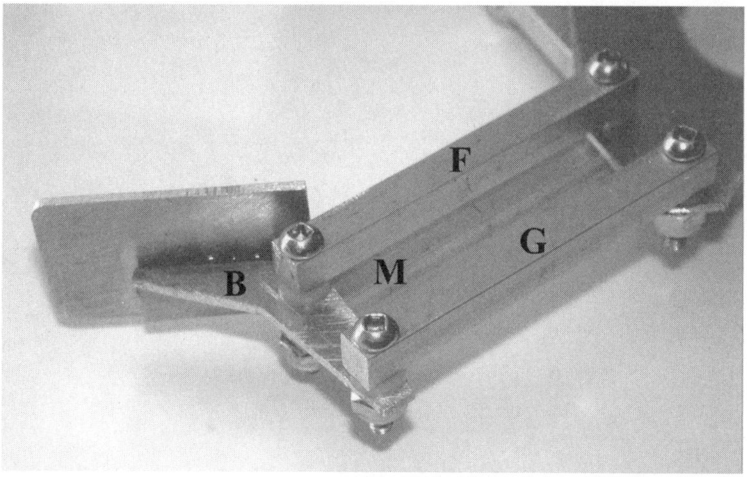

FIGURE 13.18 Gripper finger B attached to mechanical linkage pieces F, G, and M.

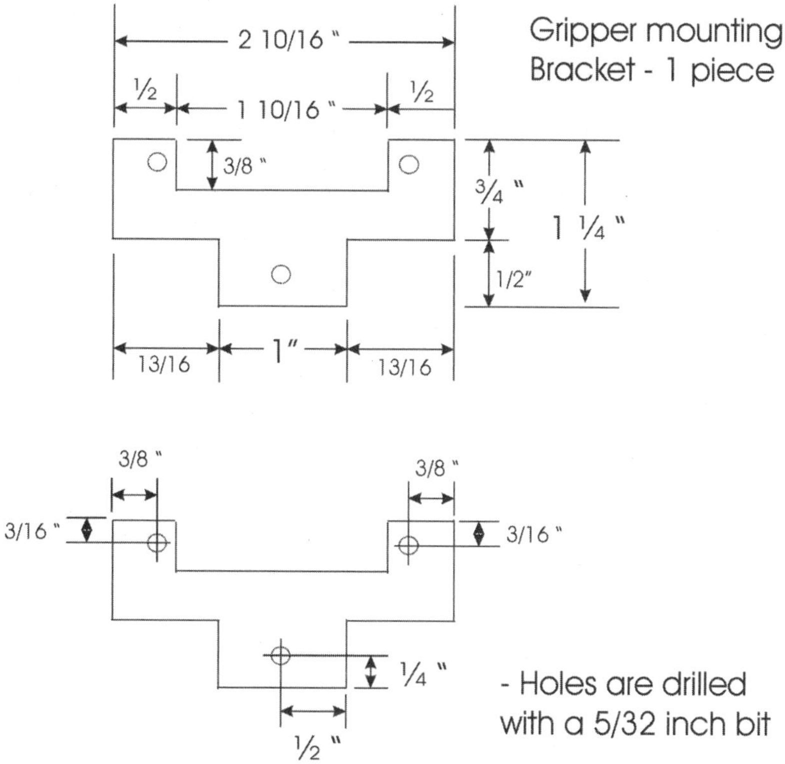

FIGURE 13.19 Cutting and drilling guide for gripper mounting bracket.

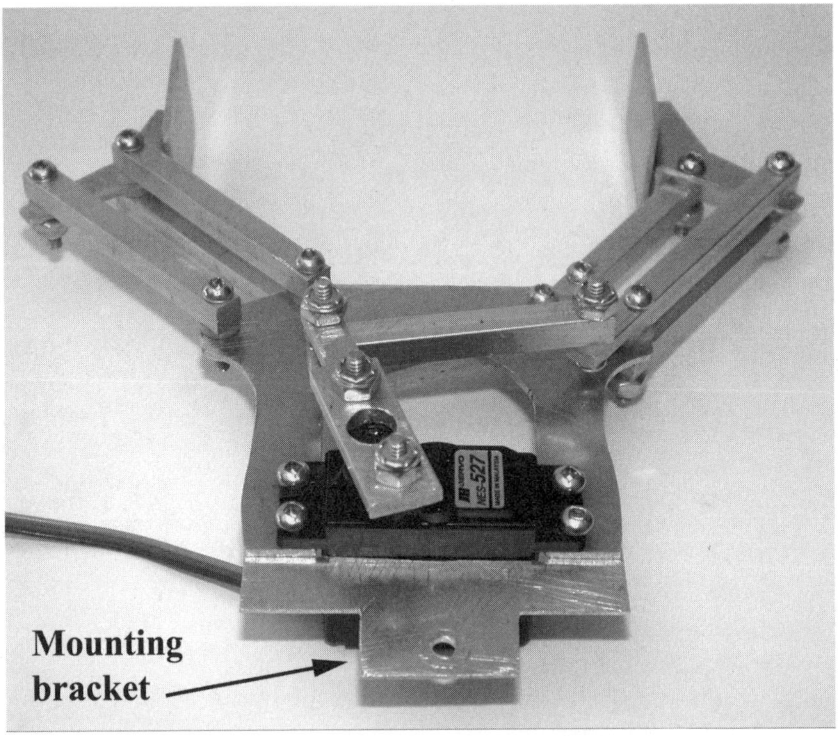

FIGURE 13.20 Mounting bracket attached to the gripper.

Remove the sonar ranger module if you haven't done so yet. The sonar connector cable can be tucked up under the circuit boards so that it is still available when you decide to reattach the ranger. To mount the gripper to the front of the robot, use the existing machine screw, washer, and nut that held the 9-volt battery clip and sonar ranger in place. Figure 13.21 shows the gripper attached to the robot.

The last thing that needs to be done to the robot's hardware is to connect the gripper servo connector to the radio receiver. String the servo connector wire under the robot and plug it into the radio receiver channel that is marked "Thro" (throttle). The radio transmitter control stick on the left will control the gripper, as shown in Figure 13.22.

Software

The software to control the robot with the gripper attached is essentially the same as the last program to control the robot with the remote control. It is called

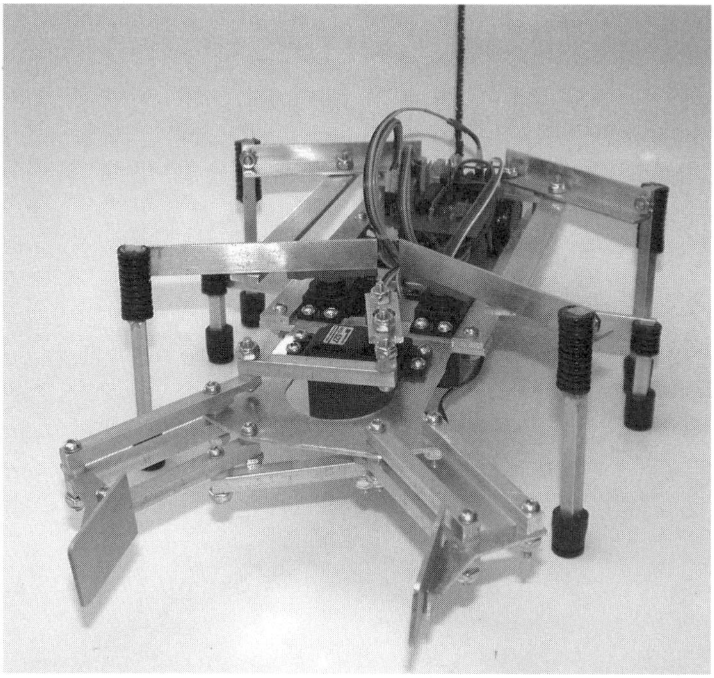

FIGURE 13.21 Gripper attached to the robot.

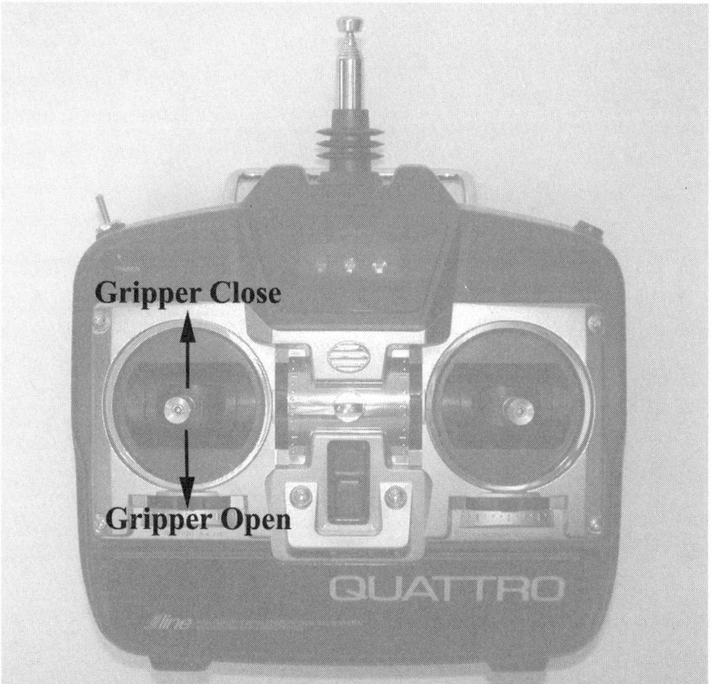

FIGURE 13.22 Radio transmitter stick positions for gripper control.

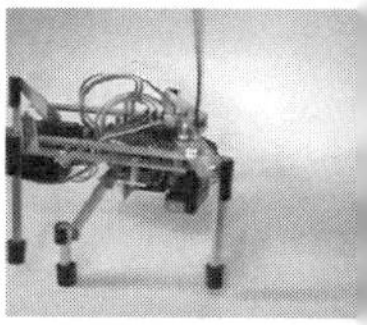

gripper-remote.bas, and is listed in Program 13.1. The corresponding gripper-remote.hex file is listed in Program 13.2. The only difference is that the code to control the sonar ranger module has been taken out since the device was removed to make room for the gripper. The only caveat is that since the robot can no longer sense its surroundings, the operator cannot be warned of any obstacles nearby, nor can the robot be programmed to avoid obstacles.

PROGRAM 13.1 **Gripper-remote.bas program listing.**

```
'******************************************************************
' gripper-remote.bas
' Robot remote control program with gripper.
' PicBasic Pro compiler - MicroEngineering Labs.
'******************************************************************

trisa = %00000011                  ' set porta pins 0&1 to inputs
trisb = %00001000                  ' set portb pin 3 to input

m_servo       var    byte          ' define variables and constants
l_servo       var    byte
r_servo       var    byte
timer         var    byte
right_led     var    PORTB.0
left_led      var    PORTB.1
piezo         var    PORTB.4
right_servo   var    PORTB.5
left_servo    var    PORTB.6
mid_servo     var    PORTB.7
chan_1        var    PORTA.0
chan_2        var    PORTA.1
Channel_1     var    byte
Channel_2     var    byte

low left_led                                 ' turn off left LED
low right_led                                ' turn off right LED

sound piezo,[100,10,50,5,70,10,50,2]         ' Make startup sound

trans_power:

     pulsin chan_1,1,channel_1               ' check to see if the
                                             ' transmitter is
     pulsin chan_2,1,channel_2               ' turned on or not
```

```
    if (channel_1 > 155 or channel_1 < 145) or (channel_2 > 155
     or channel_2 < 145) then
       sound piezo,[100,5]
       goto trans_power
    endif

main:

    pulsin chan_1,1,channel_1                       ' get receiver
                                                    ' values for
                                                    ' channels 1 & 2
    pulsin chan_2,1,channel_2

    if channel_1 = 0 or channel_2 = 0 then          ' check to see if
                                                    ' the transmitter
                                                    ' is off
       goto trans_power                             ' or out of range.
    endif

    If channel_2 > 180 and (channel_1 > 145 and channel_1 < 155)
     then
       gosub walk_forward
    endif

    If channel_2 < 120 and (channel_1 > 145 and channel_1 < 155)
     then
       gosub walk_reverse
    endif

    If channel_1 < 120 and (channel_2 > 145 and channel_2 < 155)
     then
       gosub turn_right
    endif

    If channel_1 > 180 and (channel_2 > 145 and channel_2 < 155)
     then
       gosub turn_left
    endif

    If channel_1 < 120 and channel_2 > 180 then
       gosub right_forward
    endif

    If channel_1 > 180 and channel_2 > 180 then
       gosub left_forward
    endif
```

```
    If channel_1 < 120 and channel_2 < 120 then
        gosub right_reverse
     endif

    If channel_1 > 180 and channel_2 < 120 then
        gosub left_reverse
     endif

Goto main

walk_forward:                          ' walk forward subroutine
    m_servo = 170
    gosub servo
    l_servo = 160
    r_servo = 160
    gosub servo
    m_servo = 100
    gosub servo
    l_servo = 120
    r_servo = 120
    gosub servo
return

walk_reverse:                        ' walk reverse subroutine
    m_servo = 170
    gosub servo
    l_servo = 120
    r_servo = 120
    gosub servo
    m_servo = 100
    gosub servo
    l_servo = 160
    r_servo = 160
    gosub servo
return

turn_left:                           ' rotate the robot left 18 degree
                    m_servo = 170
                    gosub servo
                    l_servo = 120
                    r_servo = 160
                    gosub servo
                    m_servo = 100
                    gosub servo
                    l_servo = 160
                    r_servo = 120
                    gosub servo
```

```
return

turn_right:                         ' rotate the robot right 18 degrees
                  m_servo = 170
                  gosub servo
                  l_servo = 160
                  r_servo = 120
                  gosub servo
                  m_servo = 100
                  gosub servo
                  l_servo = 120
                  r_servo = 160
                  gosub servo
return

right_forward:                      ' right-forward subroutine
    m_servo = 170
    gosub servo
    l_servo = 160
    r_servo = 140
    gosub servo
    m_servo = 100
    gosub servo
    l_servo = 120
    r_servo = 130
    gosub servo
return

left_forward:                        ' left-forward subroutine
    m_servo = 170
    gosub servo
    l_servo = 140
    r_servo = 160
    gosub servo
    m_servo = 100
    gosub servo
    l_servo = 130
    r_servo = 120
    gosub servo
return

right_reverse:                         ' right-reverse subroutine
    m_servo = 170
    gosub servo
    l_servo = 130
    r_servo = 120
    gosub servo
```

```
    m_servo = 100
    gosub servo
    l_servo = 140
    r_servo = 160
    gosub servo
return

left_reverse:                                  ' left-reverse subroutine
    m_servo = 170
    gosub servo
    l_servo = 120
    r_servo = 130
    gosub servo
    m_servo = 100
    gosub servo
    l_servo = 160
    r_servo = 140
    gosub servo
return

servo:                                         ' subroutine to set servos
        for timer = 1 to 10
        pulsout mid_servo,m_servo
        pulsout left_servo,l_servo
        pulsout right_servo,r_servo
        pause 13
        next timer
return

End
```

PROGRAM 13.2 Gripper-remote.hex file listing.

```
:10000000C328A8008417800484138E010C1C8E0062
:10001000232003190BE2823200319BE282320BE282D
:10002000A80059200C080D040319BE28B820841319
:100030002808800664001C281D288C0A03198D0FCF
:100040001A288006BE2828088E0601308C008D01F3
:1000500000000828050E06031D08008C0A03198D0FE1
:10006000282808008F002A08840028095A208413B1
:100070008F080319BE28F03091000E0880389000D8
:10008000F030910303199100031 98F030319BE285F
:1000900049285D2003010C1828088E1F28088E08A7
:1000A00003190301900F562880063D2857280000A9
:1000B0004028FF3A84178005BE280D080C04031958
:1000C0008C0A80300C1A8D060C198D068C188D0642
```

```
:1000D0000D0D8C0D8D0DBE288F018E00FF308E070B
:1000E000031C8F07031CBE2803308D00DF307A20ED
:1000F0006E288D01E83E8C008D09FC30031C83289E
:100100008C07031880288C0764008D0F80280C183A
:1001100089288C1C8D2800008D2808008D018F01F6
:100120008E0002309D288D018F018E0001309D28A8
:100130008D018F018E0004309D2894000F080D0260
:10014000031DA4280E080C02043003180130031903
:1001500002301405031DFF30BE280038031DFF3098
:100160000405031DFF30BE280404031DFF30BE2814
:100170008C098D098C0A03198D0A08008313031357
:10018000831264000800831603308500083086005F
:1001900083128610831686108312061083160610AB
:1001A00006308312AA001030A80064308E000A3096
:1001B000322032308E000530322046308E000A3038
:1001C000322032308E000230322001308C00053077
:1001D0008400013001200C08AC0001308C00053097
:1001E0008400023001200C08AD002C088C009B30EC
:1001F00093209E002C088C0091309820A0001E08AF
:1002000084002008B420A000A1002D088C009B30A1
:100210009320A2002D088C0091309820A400220881
:1002200084002408B420A400A50020082104840030
:1002300024082504B420A400A50064002408250493
:100240003192B290630AA001030A80064308E0054
:1002500005303220E52801308C0005308400013063
:1002600001200C08AC0001308C0005308400023005
:1002700001200C08AD002C088C0000308E209E0060
:100280002D088C0000308E20A0001E08840020085D
:10029000B420A000A1006400200821040319512902
:1002A000E5282D088C00B43093209E002C088C008B
:1002B00091309320A0002C088C009B309820A20045
:1002C000200884002208AD20A200A3001E0884009C
:1002D00022082304AD20A200A30064002208230406
:1002E0000319732932222D088C00783098209E0043
:1002F0002C088C0091309320A0002C088C009B309F
:100300009820A200200884002208AD20A200A300AB
:100310001E08840022082304AD20A200A30064006C
:10032000220823040319952943222C088C007830D5
:1003300098209E002D088C0091309320A0002D085D
:100340008C009B309820A200200884002208AD2059
:10035000A200A3001E08840022082304AD20A200EE
:10036000A3006400220823040319B72965222C087E
:100370008C00B43093209E002D088C009130932087
:10038000A0002D088C009B309820A200200884003B
:100390002208AD20A200A3001E0884002208230426
:1003A000AD20A200A3006400220823040319D92968
:1003B00054222C088C00783098209E002D088C0048
```

```
:1003C000B4309320A0001E0884002008AD20A000B7
:1003D000A1006400200821040319EF2976222C08CB
:1003E0008C00B43093209E002D088C00B4309320F4
:1003F000A0001E0884002008AD20A000A100640019
:10040000200821040319052A87222C088C00783043
:100410009820 9E002D088C0078309820A0001E089F
:10042000084002008AD20A000A10064002008210461
:1004300003191B2A98222C088C00B43093209E00AC
:100440002D088C0078309820A0001E088400200819
:10045000AD20A000A10064002008210403 19312A66
:10046000A9222B29AA30AF00BA22A030AE00A030BA
:10047000B000BA226430AF00BA227830AE007830D3
:10048000B000BA220800AA30AF00BA227830AE001D
:100490007830B000BA226430AF00BA22A030AE008B
:1004A000A030B000BA220800AA30AF00BA227830DB
:1004B000AE00A030B000BA226430AF00BA22A03043
:1004C000AE007830B000BA220800AA30AF00BA22DD
:1004D000A030AE007830B000BA226430AF00BA224B
:1004E0007830AE00A030B000BA220800AA30AF00C9
:1004F000BA22A030AE008C30B000BA226430AF0017
:10050000BA227830AE008230B000BA220800AA3099
:10051000AF00BA228C30AE00A030B000BA226430F6
:10052000AF00BA228230AE007830B000BA220800A4
:10053000AA30AF00BA228230AE007830B000BA22C2
:100540006430AF00BA228C30AE00A030B000BA22C6
:100550000800AA30AF00BA227830AE008230B00076
:10056000BA226430AF00BA22A030AE008C30B000A6
:10057000BA2208000130B10064000B3031020318C8
:10058000DA2A2F088C008D010630840080301 0207C
:100590002E088C008D0106308400403010203 00879
:1005A0008C008D0106308400203010200D306C202E
:0A05B000B10FBC2A08006300DB2A2B
:02400E00F53F7C
:00000001FF
```

Now that the robot can be remotely controlled and has the ability to grasp objects, it can be a lot of fun to send it out into the neighborhood to explore. One improvement that can be made to the gripper is to add rubber or foam pads to the fingers for a better grip.

When the robot is walking with the extra weight of the gripper, you may notice that the middle servo does not lift the robot's legs very high off the ground. This is fine if you plan on walking the robot on relatively smooth surfaces. But if you want great performance, the middle servo will need to be upgraded to one with more torque. Higher-torque servos are more expensive, but you will notice quite a difference in performance and may want to upgrade all the servos. If you plan on adding more devices to the robot, such as a video camera, then this will be required.

Another solution to the weight problem is to simplify the gripper design so that it does not use parallel linkages. Examine the movement of parts I, J, and M when the gripper is in motion. A smaller design could use that linkage concept and be driven directly from the servo. Another of my gripper designs built around a standard servo is shown in Figure 13.23. One of the gripper fingers is driven directly from the servo to eliminate having to use a mechanical linkage. Although the fingers do not remain parallel with this design, it has an amazingly large grasp of 10 inches, as shown in Figure 13.24.

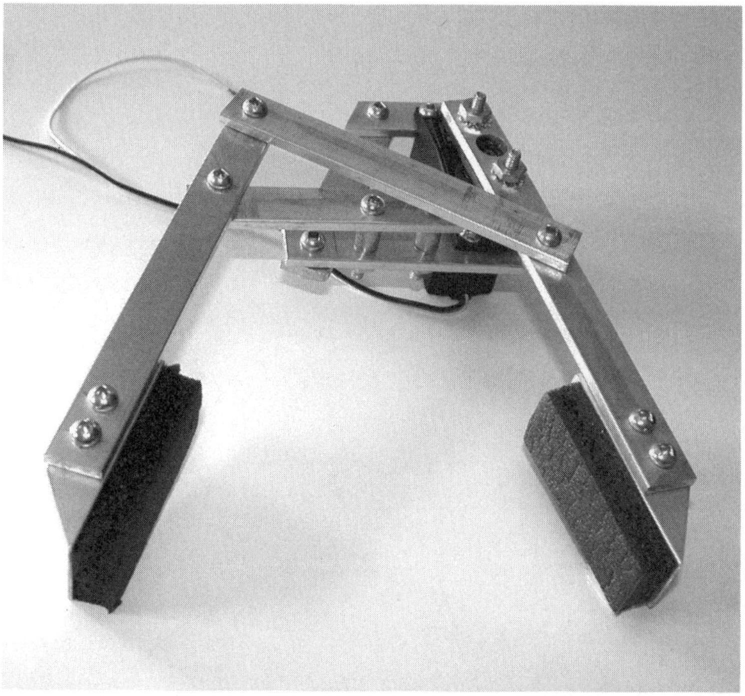

FIGURE 13.23 **An alternative gripper design built around a standard servo.**

You can experiment with interfacing the gripper directly to the controller board and adding a direct contact sensor (microswitch) on the gripper base in between the two fingers, as shown in Figure 13.25.

When the robot approaches an object that fits between the gripper fingers it would trigger the microswitch and signal the microcontroller to close up the gripper.

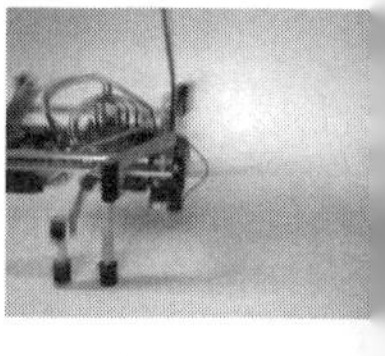

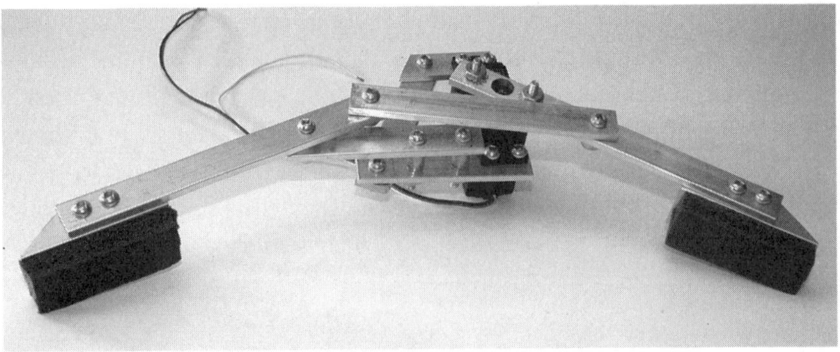

FIGURE 13.24 Gripper displaying a 10-inch grasp.

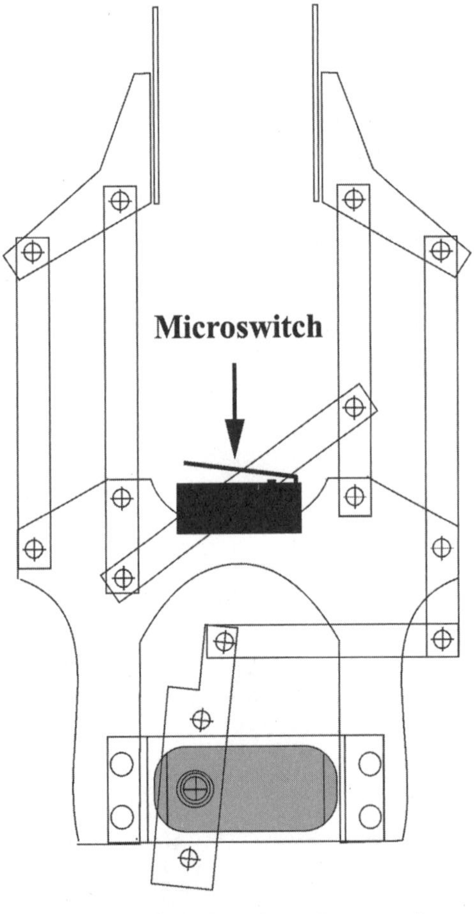

FIGURE 13.25 Position where a microswitch can be added to the gripper.

Summary

This chapter explored the design of a gripper and its addition to the robot under human control with the use of a radio. To make this design more versatile, another servo can be added to raise and lower the gripper by taking advantage of the fourth unused radio channel. The next chapter will discuss many more ideas that can be implemented with a six-legged walking robot platform.

14

TAKING IT FURTHER

There are a number of commercial, military, and humanitarian applications for small, intelligent walking robots. Because of this increasing demand, robots are now being accepted in everyday life, and are depicted in the media as a fascinating new technology that holds the promise of enhancing human existence. Following is a list of some of the experiments and extras that can be developed from the basic robot design.

1. Add a wireless data link so that a group of robots can communicate back to a central command computer and with each other. Interface a commercially available data transmitter and receiver, such as the linx transmitter and receiver available online at http://www.linxtechnologies.com/. This transmitter and receiver combination is great for sending and receiving serial information, and is very easy to implement.

   Using this technology, you could create a group of robots that work together as a swarm. Figure 14.1 shows a group of insect robots gathering and sharing information. One main objective could be determined by a human and initiated at a central computer. The goal would be broken down into smaller tasks, and assigned to each of the individual robots. The robots would work together as collective autonomous agents to achieve the common goals, much like an ant

colony. The robots could communicate their current positions and operational status with each other and the central command computer via the wireless data links. Another method would be to use short-range infrared for communications so that the messages could be relayed to the robots that are close to each other without the threat of having the radio transmissions jammed by hostile entities.

A swarm of small, covert walking robots would be suited for reconnaissance missions into unfriendly or inaccessible territory.

FIGURE 14.1 A swarm of insect-like robots working as a collective.

2. Sensors that can measure temperature and humidity can be added so that readings can be taken at different remote locations and the information radioed back to a main computer or stored in the robot's memory to be retrieved at a later date. Write an algorithm to account for differences in temperature, and the effect it has on the speed of sound through air, since this will influence the distance measurements taken with the sonar ranger. Send the robot on a simulated mission to Mars in the backyard.
3. Incorporate a gyroscope or digital compass into the control system so that the robot can keep its bearing when it is commanded to walk in a straight line. Interfacing a gyro was a difficult task in the past. It has been made easy now that cheap, lightweight gyros have been developed for radio-controlled helicopters.

These devices can be purchased at most R/C hobby shops or online at http://www.horizonhobby.com/. Figure 14.2 shows a helicopter gyroscope manufactured by JR Remote Control.

FIGURE 14.2 **The JR G410T piezo gyro—small and inexpensive.**

4. Add a miniature video/audio camera and transmitter for remote visual operation. The remote control that was added in Chapter 12 has four channels. The extra two channels could be used to pan and tilt a small video camera attached to the robot, like the one shown in Figure 14.3. Miniature video cameras and transmitters are available online at http://www.supercircuits.com/.

FIGURE 14.3 **Inexpensive, miniature video camera.**

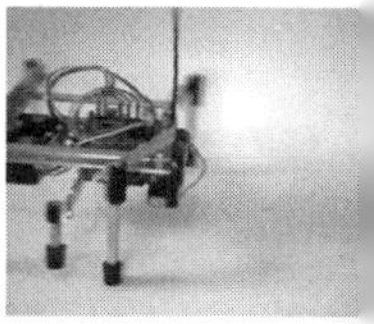

5. Attach a metal detector to the front of the robot, and have it grid-search high-traffic areas of a beach during the evening, when all the people have gone home. You will probably have to do this project with the gyroscope or compass already onboard. Have the robot mark the areas in the sand where it has detected metallic objects by rotating 360 degrees so that a definite circular track is left. Once the robot has marked the area, it can continue the search.
6. Interface a global positioning system (GPS) module to the Picmicro MCU and have the robot move from one defined area to another. Hold a competition where robots have to find their way from one position in a large city like Toronto to another position completely autonomously. The robots would have to rely on the GPS for their positional coordinates and deal with traffic, sewers, humans, and all other potentially fatal hazards. The robots would have to charge their batteries from time to time, and redefine their walking gaits and behaviors based on the injuries they have sustained during their travels. The robots would be released, and then it could take days or weeks for the robots to achieve their goals.
7. Build a dual-tone, multifrequency (DTMF) signal decoder circuit and use a cell phone to control the robot and other devices. The robot could be used as a sentry to prowl an area, checking for intruders by monitoring for human body heat using an infrared sensor. The robot could call you on the phone when any breaches to security occur.
8. Attach a small vacuum to the front or back of the robot and have the robot run a wall-following routine in rooms that need to be cleaned. Another idea is to incorporate two light sensors underneath the front of the robot, and have it follow a white line.

Summary

What has been discussed in this book has been a document of my real-life experimentation with a walking robot. I hope it will inspire you to develop your own ideas further. Take what you want from the book, and expand on the existing designs where you see fit. Discover the personal satisfaction of inventing and creating artificial life forms! I am always amazed to see a new robot come to life for the first time.

BIBLIOGRAPHY

Myke Predko, *Programming and Customizing the PIC Microcontroller*, McGraw-Hill, New York, 1998, ISBN 0-07-913645-1

John Iovine, *PIC Microcontroller Project Book*, McGraw-Hill, New York, 2000, ISBN 0-07-135479-4

D J Todd, *Walking Machines, An Introduction to Legged Robots*, Anchor Press, Great Britain, 1985, ISBN 0-85038-932-1

Barbara Webb, Thomas Consi, *Biorobotics, Methods and Applications*, MIT Press, Massachusetts, 2001, ISBN 0-262-73141-X

Anita M. Flynn, Joseph L. Jones, *Mobile Robots, Inspiration To Implementation*, A K Peters, Massachusetts, 1993, ISBN 1-56881-011-3

Stan Franklin, *Artificial Minds*, MIT Press, Massachusetts, 1997, ISBN 0-262-56109-3

Newton C. Braga, *Robotics, Mechatronics, and Artificial Intelligence*, Newnes, Boston, 2002, ISBN 0-7506-7389-3

Gordon McComb, *The Robot Builder's Bonaza*, 2d ed., McGraw-Hill, New York, 2001, ISBN 0-07-136296-7

Rodney Brooks, *Flesh and Machines*, Random House, New York, 2002, ISBN 0-375-42079-7

Tod Loofbourrow, *How to Build a Computer Controlled Robot*, Hayden Book Company, New Jersey, 1978, ISBN 0-8104-5681-8

John Iovine, *Robots, Androids, and Animatrons*, 2d ed., McGraw-Hill, New York, 2002, ISBN 0-07-137683-6

David L. Heiserman, *Robot Intelligence with Experiments*, Tab Books, PA, 1981, ISBN 0-8306-9685-7

Mark E. Rosheim, *Robot Evolution, The Development of Anthrobotics*, John Wiley & Sons, New York, 1994, ISBN 0-471-02622-0

Geoff Simons, *Robots, The Quest for Living Machines*, Sterling Publishing, New York, 1992, ISBN 0-304-34414-1

Steven Levy, *Artificial Life*, Random House, New York, 1992, ISBN 0-679-74389-8

INDEX

Note: Boldface numbers indicate illustrations and tables.

ABOUT THE AUTHOR

Karl Williams is currently employed by Mitra Imaging, a leading medical imaging software company recently acquired by AGFA. In 1985, Mr. Williams was the recipient of an IBM computer technology award for building a computer-controlled robotic arm. A resident of Ontario, Canada, he hosts a robotics and electronics Web site.